Klaus Eichmann

# Köhler's Invention

**Birkhäuser Verlag**
**Basel · Boston · Berlin**

Prof. Dr. Klaus Eichmann
Max-Planck-Institut für Immunbiologie
Stübeweg 51
D-79108 Freiburg

**Library of Congress Cataloging-in-Publication Data**
Eichmann, Klaus, 1939–
Köhler's invention / Klaus Eichmann.
p. cm.
Includes bibliographical references.
ISBN-13: 978-3-7643-7173-9 (alk. paper)
ISBN-10: 3-7643-7173-0 (alk. paper)
1. Köhler, Georges. 2. Immunologists--Germany--Biography. 3. Hybridomas.
4. Monoclonal antibodies. I. Title.
QR180.72.K64E33 2005
610'.92--dc22
[B] 2005048131

**Bibliographic information published by Die Deutsche Bibliothek**
Die Deutsche Bibliothek lists this publication in the Deutsche Nationalbibliografie; detailed bibliographic data is available in the internet at http://dnb.ddb.de

ISBN-10: 3-7643-7173-0 Birkhäuser Verlag, Basel - Boston - Berlin
ISBN-13: 978-3-7643-7173-9


Part of Springer Science+Business Media
Printed on acid-free paper produced from chlorine-free pulp. TFC ∞
Cover design: Micha Lotrovsky, CH-4106 Therwil, Switzerland
Cover illustration: Portrait photograph of Georges Köhler (Lore Lay, MPI Freiburg); page from Köhler's lab book (courtesy of Deutsches Museum, Bonn)
Printed in Germany
ISBN-10: 3-7643-7173-0
ISBN-13: 978-3-7643-7173-9
9 8 7 6 5 4 3 2 1 www. birkhauser.ch

# Table of contents

## Acknowledgements

Writing this book would not have been possible without the help of a number of people and institutions. First of all I am indebted to Marion Kazemi and Eckhard Henning of the Archive of the Max Planck Society, Berlin. They placed the entire collection of Georges Köhler's papers, notebooks, letters, etc., as well as the Köhler files of the Biomedical Section, at my disposal. I am grateful to Sophie Kratsch-Lange and Andrea Niehaus of the German Museum, Bonn, for giving me access to Köhler's notebook containing the first hybridization experiment, and their files on monoclonal antibodies and on Köhler's membership in their Board of Advisors. I have greatly benefited from casual information about Georges which I received informally from numerous mutual colleagues and friends, to whom I express my gratitude collectively. I had the privilege of more extended interviews with Rainer Hertel, Hanjörg Just, Rolf Kemler, Fritz Melchers, Celia Milstein, Michael Reth, David Secher, Davor Solter, and I am deeply grateful to all of them for openly sharing their recollections with me. I respect the wish of Claudia Köhler to not have her personal recollections about her husband, and that of her children about their father, disclosed to the public.

My sincere thanks go to Lore Lay, Nele Leibrock, and Celia Milstein, for letting me choose from their collection of photographs. The reproductions of original materials provided by the Construction Department of The Max Planck Society and by the German Museum are gratefully achnowledged. In this book I quoted from original scientific literature of many fellow scientists, some of which I know rather well, and included some original figures, with their kind permission. Nevertheless, the accounts given in this book on the evolution of the antibody problem and of the history of the hybridoma technique is a very personal one and not to be taken as a scientifc review. It is influenced by the attempt I was making to write for a general audience, and by my personal memories and preferences. It required presenting a selective rather than a complete account of the history of the antibody problem and the hybridoma technique, and selection unavoidably results in personally biased points of view. I apologize to all of my colleagues whose work I have neglected to quote, or have quoted in ways that they may not approve of. Last but not least, I am deeply grateful to my editor, Beatrice Menz. Without her continuous encouragement this book would have never been completed.

K. Eichmann
March, 2005

# Part I

# The time before

## Chapter 1

# A short history of the antibody problem

The invention of monoclonal antibodies by Georges Köhler and Cesar Milstein was made as part of an effort to generate better tools for studies of the antibody problem, and decidedly not with any putative aim to create better antibodies for scientists to use as reagents in experiments or for medical purposes such as diagnostics and therapy. It is not untypical for a scientific invention to have consequences that have not at all been forseen or planned by the inventors, or others involved in the process such as agencies that provide funds for the research. It is rather an exception in science that a problem, such as for example the treatment of a disease, is solved by scientists who plan to tackle this specific problem, and apply for and receive funds to do so. More often solutions come from rather unexpected angles. It is therefore informative to discuss here the process by which the hybridoma technology was invented, how it ultimately failed to serve the purpose anticipated by Köhler, and how it became the basis of a totally unforeseen revolution in science and medicine. Administrators, donors, and politicians in charge of managing the financing of research laboratories are well advised to appreciate that the creative process of research is to a large extent governed by adventure and randomness and, though it often leads to a goal, it rarely leads to that initially aimed at.

The antibody problem, the incentive of Köhlers's invention, has been in the center of the science of immunology since nearly its beginning at the end of the 19th century, and is still not entirely solved at the start of the 21st century. Although many generations of highly intelligent scientists were determined to tackle the problem during their time, it took an entire century to accumulate sufficient knowledge, the sum of which finally unravelled most of the story. It took a large number of research efforts, each one building on top of many previous observations made by many groups of researchers over a period of more than a human lifetime. There were many attempts and approaches that led nowhere or even misled investigators for some time. Here I can discuss only the main avenues of research that finally led to the truth.

Antibodies were first discovered in the last decade of the 19th century by the German Emil von Behring together with his Japanese co-worker Shibanasaru Kitasato. It has previously been known that animals that had survived an infection with, say, the causative agent of whooping cough, *Bordetella pertussis*, were thereafter resistant to reinfection with the same bacteria, but not to infection with another bacterium, say, the causative agent of diphtheria, *Corynebacterium diphtheriae*. Von Behring and Kitasato were primarily interested in the toxins of bacteria and discovered that blood fluid from animals pretreated with bacterial extracts could transfer resistance to the toxins of the bacteria onto another animal that had no previous history of having been exposed to the bacteria. They concluded that pretreated animals produced what they called "antitoxins" that were circulating in the blood. The blood fluid (serum) from the immune animal that was injected into the second animal had to be deprived of the clotting proteins and contain no cellular components, suggesting that a soluble molecule was responsible for the transfer of immunity. Most importantly, the transferred immunity was also restricted to the agent that was used to pretreat the first animal; it was specific to the toxin of *B. pertussis*. However, if the first animal was pretreated with extracts of *C. diphtheriae*, its serum transferred immunity to the toxin of that agent but not to that of *B. pertussis*. There was perfect specificity.

In the following years von Behring and others discovered that the substance transfering immunity was a protein that firmly attached to the toxins against which it was induced, thereby somehow inactivating them. It was soon noticed that such proteins could not only attach to and neutralize toxins but also whole microorganisms. The terms antibody and antigen were coined. The antibody proteins were found in the globular fraction of the serum proteins and were therefore also referred to as immunoglobulin. Paul Ehrlich, another German scientist, was fascinated by certain dyes that were in use in microscopic pathology and that specifically stain certain types of cells or certain subcellular structures, but not others. For example, such dyes allow one to discriminate muscle cells from liver cells or the nucleus from the cytoplasm of a single cell in the microscope. Comparing the two types of specific substances, antibodies and dyes, Ehrlich realized that the ability to attach specifically to a given structure was the result of molecular complementarity. This is now known to be a general biological phenomenon. Certain pairs of natural molecules strongly attach to one another, for example enzymes and their substrates. Parts of the molecular surfaces of the antigen and of the antibody, or the enzyme and its substrate, fit together like key and lock, or like pot and lid. As a result, because of the close molecular proximity between antigen and antibody in the region of complementarity, a large number of weak molecular interactions can form between the antigen and the antibody, together resulting in rather strong attachment and neutralization of the dangerous properties of infectious agents.

Here the antibody problem starts. Soon it was realized that the number of different antigens that antibodies could specifically attach to was enormous. Most substances of natural origin including proteins, sugars, fats, and nucleic acids, could act as antigens to which antibodies specifically attach. Each of these natural molecules occurs in a sheer endless variety in different organisms in nature. Ehrlich had already shown that antibodies can be induced by immunization of rabbits with human red blood cells, a non-infectious antigen. Landsteiner, an Austrian scientist working at the Rockefeller Institute in New York, showed in the 1930s that specific antibodies could be induced in animals not only by substances of biological origin but with molecules of any nature including artificially synthesized chemicals that do not occur in nature. The diversity of antibodies should be of corresponding magnitude. Estimates of how many different antibodies a person or a laboratory animal can make increased over the years. Today we know that this number may be close to $10^{12}$ for a human being.

Why is it a problem that a single person can make that many different antibodies? In order to appreciate the existence of the antibody problem it is important to take into consideration that proteins are encoded by genes and to remember some details on how genes are constructed and work. Indeed, knowledge in genetics developed in parallel with that on antibodies and tremendously stimulated antibody research. In the 1940s, Avery, MacLeod, and McCarty, also working at the Rockefeller Institute in New York, were studying the phenomenon of bacterial transformation. A non-capsule-producing strain of the bacterium *Streptococcus pneumoniae* can be transformed to produce a capsule, simply by adding an extract of the capsule-producing strain to the culture broth. Thus, bacteria can be induced to permanently produce a novel protein, i.e., the capsule protein, for many subsequent generations. As the ability to produce the capsule protein was heritable in the transformed bacteria, the substance in the extract causing this transformation was likely to be the material responsible for inheritance. In a painstaking process of fractionating and analyzing the multiple mostly unknown components present in the bacterial extract they identified this substance as deoxyribonucleic acid (DNA). The results were published in 1944 and rank among the most important milestones of biological research in the 20th century. In the subsequent decades the molecular principles of inheritance were unravelled in great detail by many groups of investigators. DNA is a main component of chromosomes and is the substance of inheritance not only in bacteria but in all living organisms. It consists of very long and seemingly random strings of four different components, termed nucleotides or bases. The British scientists Watson and Crick found that these strings were arranged pairwise in a double helix. The four different bases make up two pairs, one partner of a pair being complementary to the other. The two partners of a pair of bases are juxtaposed in the double helix as so-called base pairs, so that one DNA strand is in its entirety complementary to the other. In higher organisms each cell contains the same two sets of chromosomes, one from the father and one from the mother, which consist of double-stranded DNA. During somatic cell division the two strands of the double helix separate and each strand serves as a template for the synthesis of a new strand in the daughter cells. This process is called DNA replication and means that one strand of DNA can be indefinitely reproduced in nature. In this way the daughter cells maintain identical genes arranged on two identical sets of chromosomes over many generations of cell division. However, stability is not absolute. Nucleotides can sometimes be exchanged by faulty replication, leading to mutations and in some cases to functional alteration of genes. The process is slightly different in germ cells. Upon division, the double set of chromosomes present in somatic cells is halved leaving sperm and egg with only a single copy of each chromosome. Upon fertilization the homologous chromosomes from egg and sperm pair up and form a new double set in the zygote, the first cell of the new embryo. In this way we inherit the genes from both of our parents.

How can DNA code for all the other molecules that constitute our body? All organisms consist of proteins, sugars, and fats, in addition to minerals and water. Out of these, only proteins are directly encoded by DNA. Fats and sugars are synthesized by enzymes which are proteins, and so these constitutents are also determined by DNA, though indirectly. Proteins consist of one or more polypeptide chains, each consisting of a string of usually between about 50–500 amino acids. Amino acids are linked to each other by the peptide bond which links the amino group of one to the carboxy group of the next in the chain. Accordingly, polypeptide chains have one free amino-terminus and one free carboxy-terminus. 20 amino acids exist in nature thus allowing a sheer endless variety of different proteins to be made. A sequence of three nucleotides in the DNA, a so-called triplet, determines one amino acid in the sequence of a polypeptide chain. A sequence of triplets in the DNA thus determines the amino acid sequence in a polypeptide chain. The relationship of DNA triplets to amino acids is called "genetic code" and applies to all living organisms, it is universal. Out of the 64 possible nucleotide triplets, about two-thirds are used to encode the string of amino acids in a protein. The rest serves for regulatory signals, for example to indicate the start and stop of a gene etc. A DNA strand thus consists of a string of many genes. The process leading from DNA to protein involves several forms of another nucleic acid, ribonucleic acid (RNA) that serve as intermediaries, and that will be discussed below. The important dogma that developed from these studies was: One gene – one polypeptide chain.

By the mid 20th century the gross molecular composition of the antibody molecule was largely understood. Gerald Edelman at the Rockefeller Institute had shown that it consists of two chains of amino acids (polypeptide chains), termed heavy and light chains because the former is about twice as long as the latter. Rodney Porter in Britain had shown that upon digestion with proteases antibodies fall into three fragments, two of which

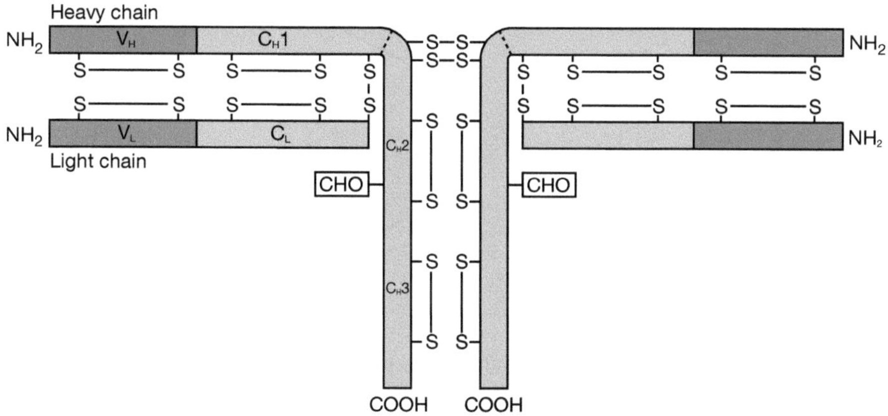


Figure 1
Schematic representation of the structure of an antibody molecule of the IgG class. It consists of two types of polypeptide chains, two heavy and two light chains, which are linked to one another by interchain disulfide bonds (-S-S-). Each chain has an amino-terminal ($NH_2$) variable (V) region and a carboxy-terminal (COOH) constant (C) region, and is composed of several so-called Ig domains of about 110 amino acids each. The V regions of either chain ($V_L$,$V_H$) and the C region of the light chain ($C_L$) consist of single domains. The C region of the heavy chain consists of 3 domains in tandem: $C_H1$, $C_H2$, $C_H3$. Each V and C domain contains an intrachain disulfide bond. The $C_H2$ domains carry carbohydrate side chains (CHO). The V domains of heavy and light chains together form the two antigen attachment sites of the molecule (from Edelman, 1970).

maintain the ability to attach to antigens. These results, for which both received a Nobel Prize, led to the well-known Y model of an antibody molecule, in which the stem is formed by the heavy chain alone whereas the two arms are formed by paired heavy and light chains. One antibody molecule has two antigen attachment sites located on the two arms. In addition, several different types of heavy and light chains could be distinguished. These led to the distinction of a handful of immunoglobulin (Ig) classes such as IgM, IgG, IgA, IgE, etc., which serve different roles in immune defense mechanisms.

So how do antibody molecules with specificity for different antigens differ from one another? In order to account for the seemingly endless variety of antibody specificities there were two opposing theories at the time, the instructional theory and the natural selection theory. Proponents of the instructional theory, including Felix Haurowitz and the two-times Noble Laureate Linus Pauling, suggested that all antibodies had the same amino acid sequence. They acquired specificity for different antigens by folding their polypeptide chains around the antigen, maintained this fold and thus became permanently instructed to attach to their antigen. The opposing natural selection theory, proposed by the Dane Niels Jerne and later extended as clonal selection theory by the Australian McFarlane Burnett,

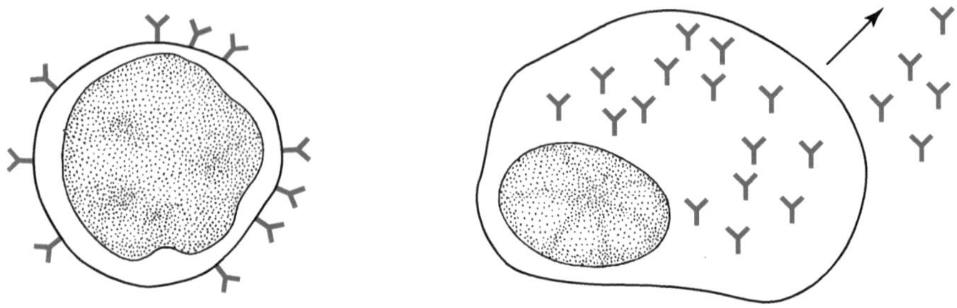

Figure 2
Antibodies are produced by B cells and their progeny, plasma cells. Before exposure to a foreign antigen, B cells are small round cells that do not divide (left). They expose their antibody molecules at the cell surface as receptors. When an antigen attaches to these receptor antibodies, the B cell is activated, begins to divide and increases in size, owing to a dramatic increase in antibody synthesis in the cytoplasm. It is now termed plasma cell (right). Antibody molecules are secreted by plasma cells and circulate in the blood. Plasma cells sometimes turn malignant and form tumors, so-called plasmocytomas or myelomas, which secrete large amounts of a single type of antibody into the blood, the myeloma protein (from Jerne, 1973).

suggested that antibodies of different specificities differed by their amino acid sequence in the antigen attachment site. Thus, a large array of antibodies with different antigen attachment sites would be preformed in the body and an antigen entering the body would select and stimulate the production of antibodies with fitting attachment sites during an immune response.

With the available technology neither theory was readily provable at the time. It was possible to determine the amino acid sequence of polypeptide chains but this required pure antibody preparations in which all molecules had the same sequence. Animals immunized with even the simplest antigen turned out to produce a surprising variety of different antibodies which did not lend themselves to protein sequencing due to sequence heterogeneity. It was at this point that immunologists began to feel the need for homogeneous antibody preparations. Henry Kunkel at the Rockefeller Institute studied patients suffering from multiple myeloma (also termed plasmocytoma), a cancer of plasma cells that secrete large amounts of antibody-like protein. Owing to his work and that of others it was realized that myeloma proteins were indeed antibodies. In a healthy person, antibodies are produced by plasma cells that are derived from B (for bone marrow derived) lymphocytes after stimulation with antigens. Myeloma proteins are homogeneous because they stem from a tumor of plasma cells originating from a single plasma cell turned malignant. They are monoclonal. However, as the malignant transformation hits cells at random, there was usually no antigen known for any of the myeloma proteins. Myeloma proteins were thus antibodies without known specificity and not a perfect material for studies on the molecular basis of specificity. Nevertheless, the German scientist Norbert Hilschmann, working in the laboratory of Lyman Craig at Rockefeller, reasoned that the myeloma proteins from two different patients would almost certainly differ in their unknown specificities and performed an experiment in which he determined the amino acid sequence of the light chains of two different myeloma patients. The results, published in 1965, demonstrated that the two light chains had the same sequence in the carboxy-terminal half but differed drastically from one another in the amino-terminal half. This result was a milestone in a long series of further experiments proving the natural selection theory to be correct and revealing that antibody polypeptide chains have a variable and a constant region. Antibodies of different specificity differ in amino acid sequence in the variable region of both heavy and light chains. Antibodies belonging to different classes differ in the amino acid sequence primarily of the constant region of the heavy chain. The same heavy chain variable region sequence was often found in combination with different heavy chain constant regions, i.e., with different antibody classes. The situation prompted a break of the dogma from one gene–one polypeptide to two genes–one polypeptide for antibody heavy and light chain genes, as first suggested by Dreyer and Bennet. In other words, immunologists began to think that the variable and constant regions of antibody polypeptide chains were somehow jointly encoded by two genes, a V (variable) and a C (constant) region gene. However, by collectively disproving the instructional theory these findings generated a novel puzzle. While the instructional theory caused no problem with respect to the number of genes required for antibodies, the natural selection theory required either very large numbers of inheritable antibody V-genes or mechanisms previously unheard of that increase genetic diversity of antibody genes somatically, i.e., in the very cells that produce the antibodies. While all agreed that the C-genes were inheritable, again two theories were put forward for the V-genes. The germline theory postulated that all antibody V-genes were present in the germline (nowadays called the genome) and were inherited from parents to offspring. The somatic mutation theory postulated that a few antibody V-genes were inheritable, and would diversify by mutations in the cells that produce the antibodies.

Many discussions were held on whether the mammalian genome is large enough to accomodate the number of genes necessary to encode all antibodies. If heavy and light chains were freely combinable, with each combination giving rise to a unique attachment site, one would need one million different chains of each to generate $10^{12}$ different attachment sites. This would require 2 million genes for antibodies, not counting the different classes. At this early time to have 2 millon antibody genes was doubtful but not firmly excluded. Arguments were based on considerations rather than facts. For example, using the spontaneous mutation rate of DNA in bacteria for calculations on the stability of mammalian DNA, the Japanese geneticist-philosopher Susumo Ohno suggested that mammals with more than 100,000 genes would accumulate mutations so fast that the species would be unstable. Then more and more was learned about mammalian DNA, for example that actual genes account only for a small proportion of the total DNA, the rest is repetitive and non-coding and serves structural or other unknown functions. Sometimes it is referred to as nonsense DNA. From the genome analysis we now believe that humans have somewhat less than 30,000 genes coding for proteins. It is therefore clear that antibody diversity must result from somatic diversification of antibody genes.

What are the mechanisms of somatic diversification involved in this process? All previous structural studies on antibodies had been performed on isolated antibody or myeloma proteins, and the conclusions as to the nature of antibody genes had been deduced from these protein studies. Now it became obvious that an understanding of the somatic diversification mechanisms required a direct assessment of the nucleic acids encoding the antibody polypeptide chains. Since the early 1970s biologists had learned how to determine the nucleotide sequences of DNA. One of the most surprising results from the early DNA sequencing studies was that genes of higher organisms are not continuous, they exist in pieces. The pieces, called exons, are separated by non-coding stretches of varying length, called introns. In order to understand how such discontinuous genes can code for continuous polypeptides we now have to introduce the second nucleic acid, RNA. RNA consists of strands of four different nucleotides that are complementary to the nucleotides of DNA, and is an intermediate template between DNA and the polypeptide. For protein synthesis, an RNA strand is first synthesized on the DNA template as an exact copy including both exons and introns, a process termed transcription. The RNA is then processed by appropriate enzymes that cleave at the exon/intron boundaries and splice the exon pieces together to generate a so-called messenger (m) RNA. The mRNA then serves as the immediate template for polypeptide synthesis, by assembling a string of amino acids determined by RNA triplets, a process called translation. The RNA triplets are, of course, complementary to the DNA triplets so that DNA remains the ultimate template in protein coding. The intron–exon structure of a gene could be particularly well demonstrated by so-called R-loop analysis, a technology very popular at the time revealing introns in DNA that have been spliced away in the mRNA. Upon hybridization of an mRNA probe to the DNA the introns in the DNA would not find complementary sequences in the RNA and would form single-stranded DNA loops, in contrast to the exons that would hybridize to the RNA and would form double RNA–DNA strands, discernible in the electron microscope. With this new knowledge it was all of a sudden no longer a mystery how two genes, a V-gene and a C-gene, can encode an antibody polypeptide chain. However, the two gene–one polypeptide hypothesis was less than half way to the truth. It would not be the antibody system if its genes functioned according to the common rules.

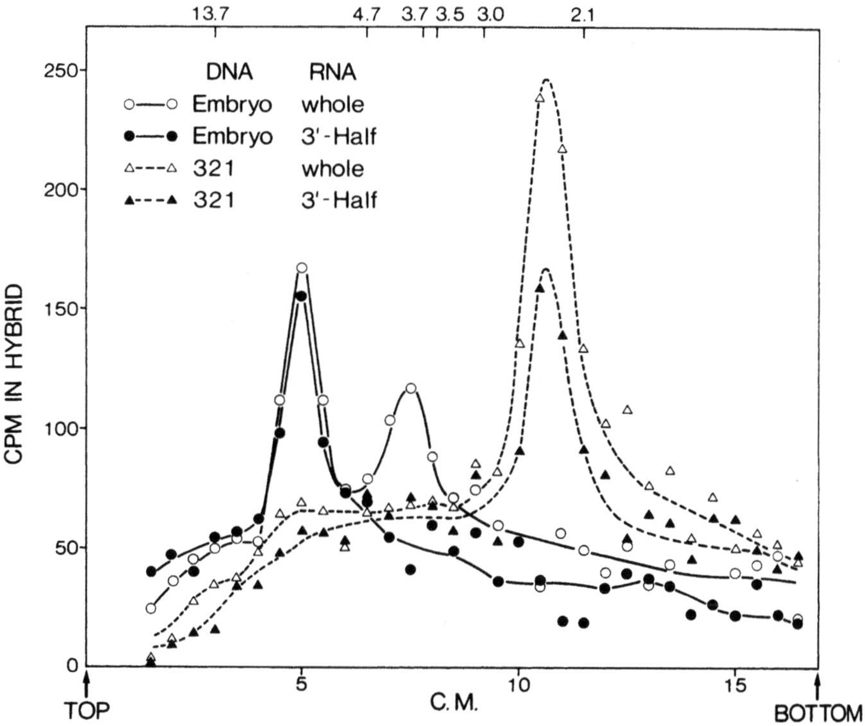


Figure 3
Tonegawa's experiment demonstrating rearrangement of antibody V and C genes in B cells. DNA was prepared from a mouse embryo and from MOPC321 myeloma cells. The DNA was fragmented with restriction enzymes, the fragments separated according to their length by vertical electrophoresis, and collected in multiple fractions from top to bottom. The fractions were tested by hybridization to radiolabeled RNA probes either corresponding to the light chain V region alone (3'half), or to the entire light chain including V and C region (whole). RNA–DNA hybridization was monitored by radioactivity (vertical axis) and plotted against the distance of migration (horizontal axis). Note that the V region RNA hybridizes to single DNA fragments which differ between embryo (closed circles) and myeloma (closed triangles) DNA. The full length RNA hybridizes to two different fragments of embryo DNA (open circles) but only to one fragment of myeloma DNA (closed circles) (from Tonegawa, 1976).

In the 1970s Susumo Tonegawa from Japan, working at the Basel Institute of Immunology, did a key experiment on antibody genes. He used the then novel technique of DNA–RNA hybridization in order to search for the V and C genes that encode the light chain of a myeloma protein. In this technique, a radioactive RNA probe is mixed with DNA which has been split up into single strands by high temperature. The RNA probe finds the DNA sequence to which it is complementary and hybridizes to it. In Tonegawa's experiment a radioactive RNA probe was prepared covering the full length of the light chain gene. The probe was then hybridized to DNA isolated from the myeloma cells and to DNA isolated from embryonic tissue before antibody-producing cells are generated. Before hybridization, the DNA was fragmented into pieces of different length that could be separated by electrophoresis in an agar slab. The pieces to which the radioactive RNA probe had hybridized, and which therefore contained the light chain genes, were then identified in a radioactivity counter. The result showed that the RNA probe hybridized to a single fragment in the myeloma DNA but to two different fragments of the embryonic DNA. This meant that in the myeloma cell the V and C genes for the light chain were close together whereas in the embryo they were far away from one another. The experiment proved for the first time that antibody V and C genes are apart from each other in the genome but change their position towards greater proximity in the cells that produce the antibody. Subsequent experiments revealed that this was also the case for the heavy chain V and C genes and characterized this process in great detail. The term "gene rearrangement" was coined. It occurs at the DNA level and is entirely different from the exon splicing on the RNA level that applies to all genes of higher organisms. Tonegawa received the Nobel prize for his discovery. Only very few genes other than antibody genes are known to rearrange at the DNA level, for example the transposons in maize.

The Tonegawa experiment proved the two gene–one polypeptide hypothesis but still did not solve the puzzle about antibody diversity and the number of V genes present in the genome. A further quantum leap towards solving the puzzle was made when Tonegawa discovered that antibody light chain V genes accounted only for amino acids 1–96 of the V regions which, however, were usually more than 100 amino acids in total length. The rest of the V region was encoded by short segments that were arranged in groups in the genome DNA. By the 1970s the techniques for analyzing nucleic acids had developed to great sophistication, aided by the possibility to isolate and replicate any mammalian gene in bacteria. Large amounts of each gene under scrutiny could thus be produced for sequencing, in contrast to studying proteins that could not be readily augmented and which always presented the investigator with a quantity problem. New results appeared with stunning rapidity. The picture that developed was that light chain V genes in the genome consist of two segments, a long one encoding about 90% of the V region and therefore still termed V, and a short J (for joining) segment. Both are joined together by rearrangement at the DNA level in antibody-producing cells. The V gene thus moves closer to the C gene but does not become adjacent to the C gene. The VJ and C genes including the intron between them are transcribed into RNA. The intron is then spliced out to form a continuous VJC mRNA. The situation is similar for the heavy chain genes, except that the V gene is made up of three segments, a long one termed V, and two short ones termed D (for diversity) and J. A continuous VDJ gene is formed by rearrangement at the DNA level. The intron between VDJ and C is thereafter spliced away at the RNA level.

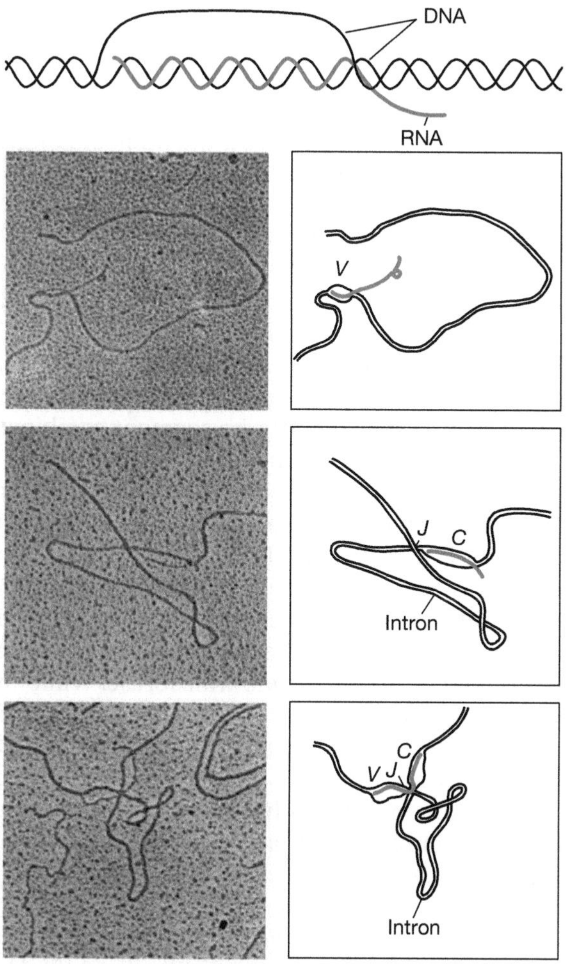


Figure 4
RNA–DNA hybridization analysis of V and C genes in embryonic and myeloma DNA. DNA is prepared from a mouse embryo and from myeloma cells, fragmented by restriction enzyme digestion, and the fragments heated to separate the two strands of DNA. The fragments are then hybridized to a full length RNA probe for the light chain under conditions which favor RNA–DNA hybridization over DNA–DNA reannealing, and analyzed by electron microscopy. The DNA stretches to which RNA has hybridized cannot reanneal and thus form loops. The top frame shows an embryonic DNA fragment to which the RNA hybridizes only by the V region, while the C region of the RNA probe remains single stranded because the C gene is on another DNA fragment. The middle frame shows an embryonic DNA fragment that hybridizes with the RNA probe by the C region. The single-stranded V region is not visible in this case. The bottom frame shows hybridization of the RNA probe to a fragment of myeloma DNA. Both V and C regions of the RNA probe hybridize to the same DNA fragment, but the long intervening stretch of double-stranded DNA shows that V and C genes are not adjacent in the myeloma DNA, only closer together. The approximate position of the J segment is indicated in the schematic drawings. From: The Genetics of Antibody Diversity, by Philip Leder 

The existence of V gene segments that can be freely combined to form somatic antibody V genes provided an elegant mechanism for somatic diversification of a relatively small number of inheritable antibody genes. But was it sufficient to generate the actual antibody diversity that can be generated? DNA sequencing soon revealed the numbers of gene segments present in the genome. While these numbers are accurate for D and J segments, they represent merely relatively accurate estimates for V segments because of their larger numbers and widespread distribution over large areas of genomic DNA. For the κ class of human light chains there are about 50 V segments and 5 J segments, giving rise to about 250 different light chain genes. For human λ light chains there are about 25 V genes and 6 J genes, resulting in 150 combinations. For human heavy chains, there are also about 100 V segments, 10 D segments and 6 J segments, giving rise to about 6,000 combinations. Consequently, by segment combination one can generate about 2,400,000 different antibodies. This is a large number but certainly falls short of the estimated figure $10^{12}$. An amplification factor of over 400,000 was not accounted for. This was disappointing as it meant that even after these major discoveries antibody diversity was not fully understood. The problem was ameliorated by findings that the joining of V gene segments was inaccurate, i.e., nucleotides could be deleted from the ends of the gene segments during joining and could be replaced at random by an enzyme termed terminal deoxynucleotide transferase. While this increased the combinatorial number of V(D)J sequences enormously, it also meant that only one out of three V(D)J genes generated maintained the correct reading frame of triplets, i.e., only 33% of the genes produced were functional. Hence, these additional diversification mechanisms could not fully account for the missing amplification factor.

Here the somatic mutations come into play. Long before knowing that antibody genes were present in segments that could be freely combined, proponents of the somatic mutation theory had already postulated that a few inheritable antibody V genes were somatically diversified by mutations. The most likely type of mutation imagined at the time were the so-called point mutations in which a single nucleotide is replaced in the DNA sequence by another nucleotide due to a mistake made in the replication process. This may alter one of the triplets in a gene so that the protein encoded by the gene may have one of its amino acids replaced by another. The new DNA strand that carries the mutation will from now on be replicated in its mutated form. Other types of mutations have also been considered, such as deletions or duplications of short DNA pieces, or gene conversion in which the gene sequence is corrected to resemble the sequence of an adjacent gene. All such types of mutations are the result of errors made during the replication process and have been observed in bacteria. In the early times of theory making, without much knowledge on nucleic acids, the existence of such mutations in mammals was merely imaginary and in most cases ambiguously deduced from protein sequencing. If such a muta-

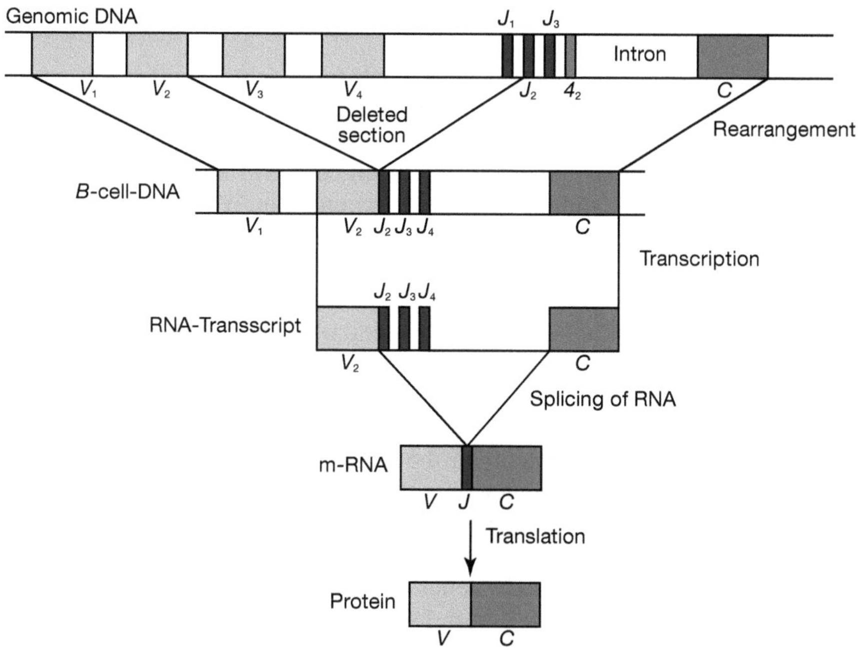


Figure 5
Schematic representation of the generation of an antibody light chain protein starting from genomic DNA, shown on top. First, in a B cell one of the V genes (in this example $V_2$) is rearranged next to one of the J segments (in this example $J_2$). Transcription generates an RNA strand that includes $V_2J_2$, the downstream J segments, the intron, and the C gene. The downstream J segments and the intron are then excised and the $V_2J_2C$ sequences are spliced together as continuous mRNA. Finally the mRNA is translated into a VC protein (from Tonegawa, 1985).

tion takes place in an antibody gene in a dividing B-lymphocyte during an immune response, a clone of daughter cells will develop that produces an antibody that differs from the original antibody by one or more amino acids. If amino acid replacements occurred in the antigen attachment site of the antibody, the specificity should be altered, i.e., either a new specificity may be generated at the expense of losing the original specificity, or the original specificity may even be improved towards a stronger attachment to the antigen. We now know that this is what indeed takes place during an immune response to an antigen. Initially, a few B cells that happen to posses antibody receptors capable of attaching to the antigen get stimulated to divide and to secrete the antibody. During cell division, mutations accumulate some of which improve the strength of attachment to the antigen, a property termed affinity. B cells that carry such mutations are stimulated even better than the original non-mutated B cells, and divide more often. The average affinity of the antibodies generated during an immune response therefore increases and the antibodies become more effective in inactivating the antigen. This process is termed affinity maturation of an immune response.

Because DNA replication is a highly faithful process, spontaneous mutations are rare. Point mutations as we know them from bacteria occur with an approximate frequency of $1/10^7$ per cell per generation. In other words, one in 10 million newly-generated bacterial cells will harbour a particular mutation, such as for example resistance to penicillin. While this seems rare, it is nevertheless one of several factors that allow bacteria to adapt to different environmental conditions. This is possible because bacteria divide every 20 min and thus generate very large number of progeny for selection. If the bacterial mutation frequency is applied to mammalian genes that possess introns, calculations on the number of mutations that a dividing B cell can accumulate and that would affect the short parts of the V genes that code for the antigen attachment site result in insignificant numbers. Thus, if somatic mutations have a role in antibody diversification, there should be a much higher mutation frequency than that of spontaneous point mutations in bacteria. Such a special mechanism should be restricted to B cells and should not operate in other cells of the body. Moreover, the mutations should be directed to antibody genes and their attachment site encoding parts, as a high mutation rate that indiscriminately affects all cellular genes would rapidly destroy genes vital for general cellular functions. In other words, there should be directed somatic hypermutation of antibody genes. This was the mechanism that Georges Köhler wanted to uncover.

Chapter 2

# The immunological scene around Köhler

Köhler entered the field of immunology in 1971 as a PhD student, joining a scientific community that was full of excitement, feeling that immunology was on its way to become the leading research activity in biomedical science. Moreover, starting in the late 1960s but with a peak in the early 1970s a big bang happened to immunology in Europe, particularly in Germany. Until that time, almost any research of consequence in this field had been performed in the United States, to some extent but by far not only by emigrants who had left Europe before and during World War II. Before the 1970s, a young European intellectual with ambition in biomedical research had to find a postdoctoral position in the US and stay for 2–3 years in order to get competent training and publish several papers. Many of the first generation post-war European postdocs stayed in the US for good. Others returned to cultivate the vast academic landscape that was in a process of happy expansion in post-war Europe. As a result, a network of centers of excellence in the field of immunology developed in European countries, together forming the breeding ground on which a talent like Köhler could prosper and eventually crystallize to become the summit of the blooming scientific development of immunology in post-war Europe.

Perhaps the most significant development in post-war European immunology was the foundation of the Basel Institute of Immunology in 1969, officially opened in 1971. This was the work of Niels Kay Jerne, a genial and charismatic leading theoretician of immunology of the 20th century. Jerne was born in 1911 to Danish parents in London. His father was a businessman in meat production, owning meat factories first in England and thereafter in Holland. Jerne went to school in Rotterdam where he excelled as an outstanding pupil. Thereafter, he seemed to have a hard time deciding what to do in terms of higher education. Initially he was unable to make decisions and worked for several periods as an employee in his father's factory, interrupted by periods of unemployment. While he loved poetry and literature, he was competent in his work and successfully installed an automated procedure for ham curing invented by his father. He then made an attempt at studying physics at the University of Leiden, but discovered his lack of excitement with this subject within four semesters which he spent mostly drinking beer with fellow students. His next choice was to study medicine at the University of Copenhagen. This he carried through to passing the MD examination and presented a thesis in 1951, at age 40, based on research carried out at the Copenhagen Statens Serum Institute on the avidity of antibodies. Antibodies remained the center of his interest ever since.

Jerne was fascinating as a person and as a scientist. His private life was described in a recent biography by Thomas Söderqvist that focusses on his chaotic love affairs. This shall not be repeated here. Suffice it to say that he was married to three different women, his first wife committing suicide. He had children with two of his spouses and with another woman to whom he was not married. As a result of going to school in several different countries he was equally fluent in four different languages, Dutch, English, German and Danish. He openly admitted that none of these languages was a mother tongue, that he had no first language of which he had full spoken and written command. He almost took pride in the fact that he did not have a feeling of national belonging. He liked to drink and could be fascinatingly charming at social events, talking a lot of nonsense, singing children's songs and reciting silly verses. On the other hand, he could be very impatient and pitilessly rude when his partner in a scientific discussion was not up to his intellectual level. I had the bad luck to be a recipient of this rudeness when he once tried unsuccessfully to explain some aspect of his network theory to me: "Klaus, you are stupid!"

Jerne clearly became the dominating figure in the big bang of European immunology. His career as an immunologist had started in the four additional years that he spent after completion of his thesis at the Statens Serum Institute. He continued to work on antibodies, studying the role that the affinity played in neutralization of bacterial toxins. In addition, he became involved in bacteriophage research, the hot subject of genetics at the time, by characterizing the neutralizing effect of antibodies to bacteriophage. This brought him together with leading geneticists including Max Delbrück, Günther Stent, and Jim Watson. Delbrück invited him to a one year research fellowship at Cal Tech, an intermezzo that eventually was not very successful and ended in a serious fallout between Jerne and Delbrück. Jerne's work on antibodies to bacteriophage laid the foundation for his revolutionary "natural selection theory of antibody formation", published in 1955 in the *Proceedings of the National Academy of Science of the USA*. He had observed that small concentrations of specific antibodies were already present in the sera of horses before they were immunized with the phage, a finding that led to the idea that millions of different antibodies were preformed in the body. A small number of such preformed antibodies would be selected by attachment to the incoming antigen and augmented to large numbers by an undefined cellular replication mechanism. Because the cel-

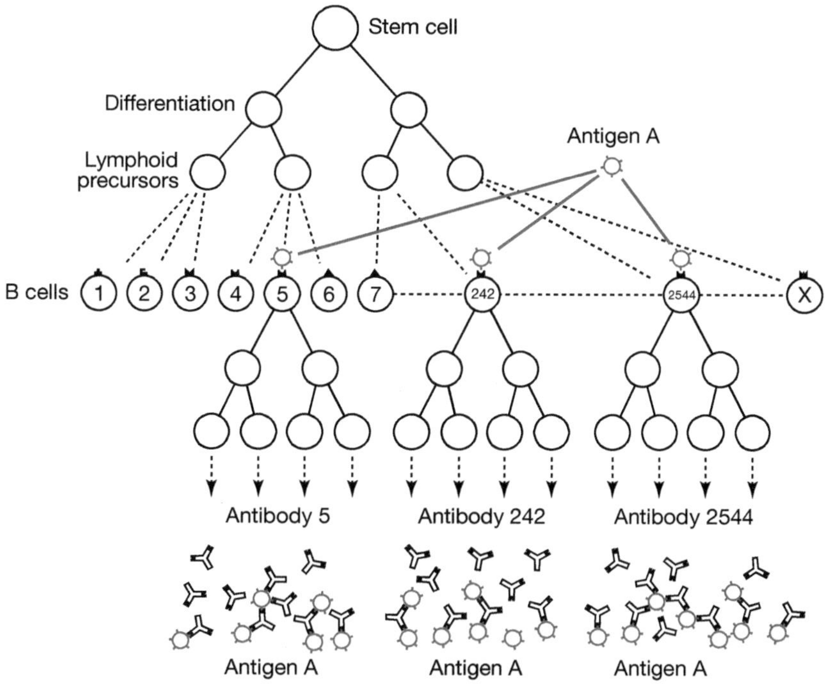


Figure 6
The clonal selection theory of the immune system. Stem cells develop into various types of committed precursors including immature B cells which rearrange their antibody genes and thus become mature B cells that expose antibody receptors on their surface, each with a different antigen binding site. Antigen A enters the body and finds receptors to which it can attach on the B cells No. 5, No. 242, and No. 2544. These B cells become activated and multiply to form clones of plasma cells secreting antibody into the blood. The antibody molecules attach to the antigen A and help eliminating it (from Edelman, 1970).

lular part remained rather vague, the theory was first vehemently rejected by the majority of immunologists who believed in the instructional theory proposed by authorities like Haurovitz, Lederberg, and Pauling. Jerne was heavily criticized not only because of the obscure cellular amplification mechanism but also because he had neglected to quote Paul Ehrlich who had proposed the concept of preformed diverse antibodies already in 1910 in his sidechain theory. The selection theory became widely accepted only much later when Macfarlane Burnet in Melbourne, Australia, published in 1957 "a modification" in which not the soluble antibody, but the antibody-producing cell bearing antibodies as receptors was the subject of selection. Selection of cells was more in-keeping with established biological principles and Gustav Nossal, an Austrian student in Burnet's laboratory, performed experiments that strongly supported the major postulate of Burnet's clonal selection theory, that one B cell produced only one antibody. In 1958

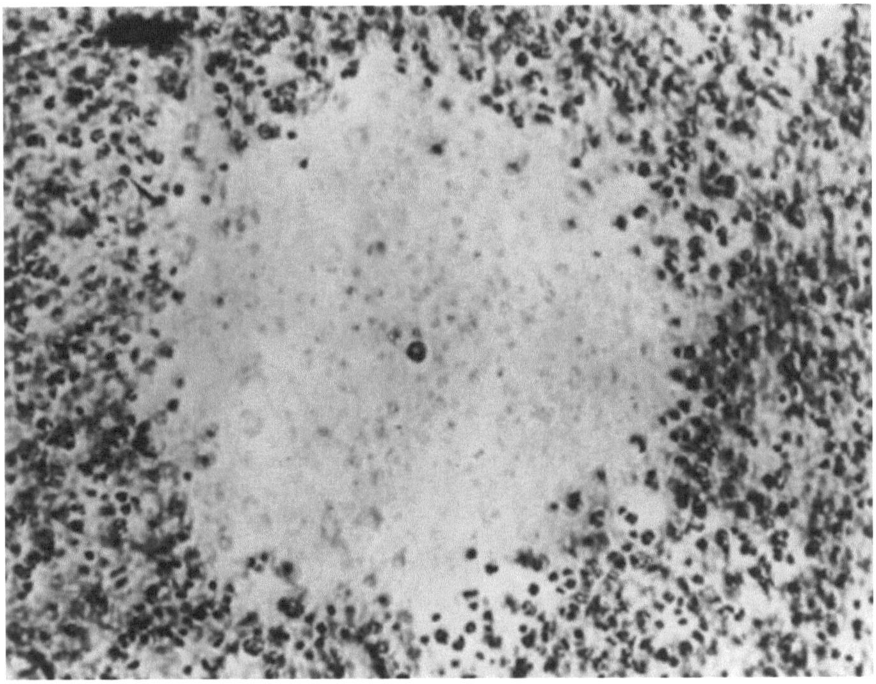

Figure 7
Jerne plaque of a plasma cell secreting antibodiy to sheep red blood cells. Spleen cells of a mouse immunized with sheep red blood cells are mixed with sheep red blood cells in warm liquid agar and the mixture poured onto a glass plate to solidify by cooling down as a thin layer. After solidification the agar is then overlayed with a source of complement and incubated for about 5 hours. During this time the plasma cells secrete antibodies that attach to the sheep red blood cells surrounding them, lysing the sheep red blood cells with the help of the complement. The area of lysis is seen as a clear halo with the plasma cell in its center (from Jerne, 1973).

Lederberg, previously a follower of instructional theories, acknowledged in a lecture that was later published in *Science* that selective theories were likely to be correct. Somewhat surprisingly, and not entirely justified, in the scientific community Jerne was later given equal credit for the clonal selection theory of antibody formation, forming the basis of his long-lasting intellectual leadership.

Jerne's work on the role of antibody affinity in toxin neutralization had highlighted the need for standardization of antibody preparations. Indeed, Jerne had been in charge of and had established methods for antibody standardization in his laboratory at the Copenhagen Institute. This had raised the interest of the World Health Organization (WHO), in charge of the international use of immunoglobulins against snake venom, tetanus and diphtheria toxins and the like, as well as of vaccines. WHO used Jerne as an advisor on standardization matters and sent him on an extended visit to several countries in Asia, where he was to inspect the local WHO outstations. Shortly after returning from Cal Tech, he took the position of head of the Section of Biological Standards at WHO, Geneva. His work there was primarily that of an international public health officer and *homo politicus*, and it stands to reason that he accepted this work as a result of his frustration with the poor acceptance of the natural selection theory. But in the wake of Burnet's biologically convincing refinement of natural to clonal selection Jerne re-emerged as a basic scientist in the beginning of the 1960s. He returned to the US to become full professor of microbiology and department chairman at the University of Pittsburgh. Here he performed the most influential practical experiment in his career, together with his co-worker Albert Nordin who had a major part in this collaborative effort. The result was the "Jerne plaque test", a method by which one could visualize individual antibody-secreting cells forming a translucent halo by lysis of red blood cells in an opaque agar plate. This assay permitted the enumeration of antibody-secreting cells and was thus a major step towards precise quantitative measurements of immune reactions. It was used all over the world for many years until simpler quantitative antibody assays became available. The original article has been cited over 4,000 times. George Köhler used this assay to detect his first hybridomas secreting antibody to sheep red blood cells.

Jerne became founding director of the Basel Institute of Immunology in 1969 after a three-year intermezzo as director of the Paul-Ehrlich Institut, Frankfurt, combined with a professorship at Frankfurt University. The Paul-Ehrlich Institut is a state laboratory working for the health ministry of the German federal government in charge of vaccine production, licensing and standardization. The bureaucratic environment of a state laboratory, which he must have experienced in similar form during his former employment with the WHO, was hardly compatible with Jerne's scientific attitude and one wonders why he took this post in the first place. According to his biographer Söderqvist, Jerne's decision to accept this post was strongly influenced by an intention to help introduce modern immunology in Germany, in addition to his wish to return to a European cultural environment. He transferred his research team from Pittsburgh to Frankfurt in 1966 and made attempts to create an exciting and creative scientific atmosphere. With the help of the newly-founded European Molecular Biology Organization (EMBO) he organized, in 1967, a much remembered postgraduate teaching course at the Ehrlich Institute at which many of the world's leading immunologists had been invited to teach their most advanced methods for the analysis of lymphocytes. However, these attempts were of no lasting avail, and Jerne realized that the Ehrlich Institute was beyond reformation. The opportunity offered by the Hoffmann-LaRoche company in Basel must have come as the ultimate relief. Jerne was given complete freedom to shape the Basel Institute according to his own philosophy. He despised hierarchies and did not permit "empire building", at least in the initial years. He personally selected the scientists he employed but thereafter they had complete freedom of chosing their subject. The financial support was generous, allowing the recruitment of promising young immunologists from all over the world. In addition, many established scientists were invited to take leave of absence from their home universities and to temporarily join the Basel Institute for the initial years. In this way Jerne succeeded in generating a sparkling intellectual environment that was entirely novel in the traditional European academic system. As a result, the Basel Institute immediately became the center of European immunology and established itself as one of the main immunological centers in the world.

Owing to his inborn talent for intellectual leadership, Jerne influenced immunological research not only at the Basel Institute but also worldwide. In the decade of his directorship in Basel Jerne published two other theories about the immune system which inspired his contemporary generation of immunologists in a similar way as had the natural selection theory. Unfortunately though, these later two theories, intelligent and fascinating as they were in their time, turned out to be rather far off the track and eventually inconsequential. The first of these, called "The somatic generation of immune recognition", was published in 1971 in the first issue of the newly-founded *European Journal of Immunology*. In this theory Jerne suggested a mechanism that would direct somatic mutations of antibody genes towards the generation of antibodies that recognize foreign but not self antigens. The theory was triggered by observations of the Swedish immunologist Morton Simonson who had found that one of about 50 lymphocytes would recognize an allogeneic transplant. This observation was a threat to the clonal selection theory that postulated a large degree of diversity of antigen-specific cells and thus a much lower frequency of cells specific for the presumably small number of antigens that differed between a transplant donor and a transplant recipient. Jerne postulated that the antibodies encoded by the V genes present in the genome of the species would recognize all the transplantation antigens, also called "major histocompatibility complex (MHC) antigens", present in that species. One individual expresses only some of the MHC antigens of the species and therefore some of its B cells would be self reactive. Such B cells are forced to mutate their V genes to escape self reactivity and thus would generate the multitude of antibodies recognizing the universe of foreign antigens. The B cells recognizing the MHC antigens not expressed in the individual, i.e., the B cells recognizing an allogeneic transplant, would not have mutated their antibody V-genes and would therefore occur at high frequency. The theory was elegant as it incorporated the problem of self–non-self discrimination, or self tolerance; an unsolved issue in immunology. How can the billions of different antibodies recognize virtually all possible foreign structures and at the same time avoid recognition of self? Jerne's second theory contained certain elements that later turned out to apply to T cells, such as the universal recognition of self major histocompatibility antigens. However, T cells do not mutate their receptor genes and, with respect to B cells, the forces that cause them to undergo somatic hypermutation, and the origin of self–non-self discrimination, Jerne was rather far off what eventually became accepted as true.

The third and last of Jerne's theories was published first in an article entitled "Towards a network theory of the immune system" in the *Annales d'Immunologie* of the Institut Pasteur in 1974, and later refined in several other publications. Jerne had become fascinated by the phenomenon that antibodies could recognize unique epitopes associated with the antigen attachment site of another antibody, the so-called idiotype (from the greek ιδιοσ = unique). This was first demonstrated by Jaques Oudin at the Pasteur Institute in Paris and subsequently became a subject of intense study in many laboratories. If a specific antibody preparation was purified and injected into another animal of the same species, antibodies were produced that recognized the idiotype of the injected antibody. Its idiotype distinguished an antibody from all others. Jerne came to the conclusion that idiotypic self-recognition among different antibodies in the immune system was unavoidable. Each individual V region would likely find a number of idiotypically complementary V regions to which it could attach. As a result a network would form including soluble antibodies and B cells bearing antibodies as receptors. If the immune system would respond to an antigen, not only would the B cells with receptors recognizing the antigen be activated, but also the B cells recognizing the idiotypes of those B cells. Idiotypic stimulation would accompany each immune response, involve large sectors of the immune system, and would serve to regulate the strength and duration of immune responses. The network theory fascinated and influenced the work of many immunologists, particularly in Europe. Over a number of years, a fair proportion of immunological research was devoted to performing experiments to understand the nature and the consequences of network regulation. Among them Klaus Rajewsky, Hans Wigzell in Sweden, Antonio Coutinho and others at the Basel Institute, and the author himself. There was an almost religious belief that went as far as considering idiotypic antibodies as vaccines against infectious diseases. In addition, it stimulated a number of mathematicians to engage in developing algorithms for system analysis of the immune network. On the other hand, there were some who doubted the scientific value of the network theory. Among them prominent colleagues such as Melvin Cohn of the Scripps Institute in LaJolla, or Göran Möller in Stockholm, who argued that a theory is worthless if it cannot be put to experimental examination. Indeed, the network theory did not make exact predictions and consequently was never really proven or disproven by experiments. For example, patients with certain autoimmune diseases get better when treated with pooled immunoglobulins, a therapy widely used until today. While this is suspected to be a network effect, a role of idiotypic recognition in this therapy has never been really demonstrated. As late as 1984, Hans Wigzell remarked in his speech at the Nobel ceremony: "*It is now well documented that network forces of the fascinating type that Jerne predicted indeed do exist inbuilt in our own immune system.*" However, as time went on, it became more and more obvious that other regulatory elements had a more prominent role in immune regulation, such as the ever increasing number of newly-discovered cytokines, receptor-mediated homeostatic proliferation or apoptotic cell death, and regulatory T cells. Idiotypic recognition did not play a role in any of these regulatory mechanisms. Today there are only a few who still believe in an important role of the idiotypic network in immune regulation. The network theory has simply faded away.

Of equal if not greater importance in Köhler's invention was the Laboratory of Molecular Biology of the Medical Research Council at the University of Cambridge, the place at which Cesar Milstein worked. Milstein was born in 1927 in Argentina to poor semi-Jewish immigrant parents who nevertheless made sure through hard work that Cesar and his brothers could go to university. In 1957 Milstein got a doctoral degree in chemistry at the University of Buenos Aires with a thesis on the enzyme aldehyde dehydrogenase. Therafter he travelled for one year through Europe and Israel with his wife, Celia, whom he had met through their mutual engagement in political student affairs. He then obtained a fellowship of the British Council and took a 3-year leave of absence from the National Institute of Microbiology in Buenos Aires to work with Malcolm Dixon in the Department of Biochemistry of the University of Cambridge, where he obtained a PhD degree presenting a thesis on metal activation of the enzyme phosphoglucomutase. He returned to Argentina only for 2 years during which he experienced the severe oppression of academic and intellectual freedom established by the ruling military junta. During his thesis in Cambridge he had met and collaborated with Fred Sanger whose Department of Molecular Biology he joined in 1963. Following the advice of Sanger, Milstein left enzymology and began to work on antibodies.

In 1962, during the short intermezzo as a junior research group leader in Argentina, Milstein developed the theory that antibodies were a random combination of short polypeptides linked together by disulfide bonds. In his Nobel lecture Milstein admitted that "*of all the prevailing theories that I am aware of, this is one that was widest off the mark.*" Nevertheless, it was this hypothesis that started Milstein's work on antibody disulfide bridges in Sanger's laboratory in 1963. He performed experiments that enabled him to determine the amino acid sequence surrounding the cysteine residues that are linked together by disulfide bridges within heavy and light chains. What he found was that antibody polypeptide chains contained constant and variable disulfide bridges. This finding fell into place with Hilschmann's discovery of the variable and constant regions, i.e., each of these regions contains a disulfide bridge. Later crystallization studies showed that constant and variable regions of antibody polypeptide chains consist of globular domains, now termed immunoglobulin (Ig) domains. An Ig domain is about 110 amino acids long, and is tightly folded into two stacked layers (so-called β sheets, as opposed to α helices) of three to four antiparallel strands of amino acids. The sheets are fixed to one another with the help of a disulfide bridge. In variable Ig domains, the loops at one end of the strands form the antigen attachment site.

Of course Milstein did not know this in 1964/5 when he puzzled over variable and constant disulfide bridges. Like many immunologists at the time, he was wondering how immunoglobulin amino acid sequence variability was generated at the nucleic acid level. Initially, Milstein intensely rejected the two gene–one polypeptide hypothesis that had just been published by Dreyer and Bennet, because he could not agree to the presence of millions of V genes in the genome. He favored somatic diversification mechanisms and temporarily became attracted by the idea of somatic gene crossovers between neighboring genes as a means to increase diversity, proposed by Oliver Smithies in 1967 and a major theme at that year's famous Cold Spring Harbour Symposium on immunology. However, the upcoming evidence of V region subgroups and their interpretation as representing a fair number but not millions of V genes, that nevertheless all shared the same C region, meant that the two genes–one polypeptide hypothesis was compatible with a limited number of V genes, and thus became acceptable also to the proponents of somatic diversification including Milstein.

Milstein was very different from Jerne. Like most other self-respecting scientists at the time, he also contributed his share of theories, for example a hypothetical gene conversion mechanism on somatic diversification published together with Sidney Brenner in 1966, or an expansion/contraction theory on the evolution of immunoglobulin genes published together with Richard Pink in 1970. However, Milstein was basically an experimentalist who firmly believed only in solid experimental facts. By 1970 Milstein was frustrated with the inconclusive nature of protein sequences and was among the first who realized that no amount of protein sequencing would ever lead to the final solution of the antibody problem. Instead, one would have to turn to the analysis of the genes themselves. DNA sequencing did not exist at the time, but in Fred Sanger's laboratory a method had been developed to sequence RNA. In addition, a method had just been described for *in vitro* protein synthesis using isolated mRNA as a template. Milstein started to isolate immunoglobulin mRNA from myeloma cells growing in tissue culture, to verify the successful purification by *in vitro* protein synthesis, and to determine the RNA nucleotide sequences. There were two results of this work. One was that V and C genes were continuous in the mRNA, indicating that the joining of the V and C regions occurred at the DNA or immature RNA but not at the protein level. The second result of this work was the unexpected discovery of the leader sequence of secreted proteins, a fundamental finding with general consequences for cell biology far beyond immunology. The question was how proteins that are synthesized in the cytoplasm of the cell can transit the lipid membrane of the secretory vesicles that eventually fuse with the outer cell membrane to spill their contents in the surrounding fluids. Milstein discovered that proteins destined for secretion, such as antibody light chains, are synthesized with an amino-terminal extension of about 20–30 amino acids. The extension is rich in hydrophobic amino acids that like to penetrate into the equally hydrophobic lipid bilayer of the vesicle membrane and thus channel the rest of the polypeptide chain into the vesicle. Once on the inside, a protease cleaves the leader sequence off.

The ultimate aim of mRNA sequencing was to discover the nature of somatic diversification of antibody V genes. To this end, mutant cells had to be generated from a wild-type cell. The experiment designed by Milstein and his collaborators David Secher and Richard Cotton relied on a mouse myeloma originally generated by Michael Potter, and termed MOPC (mineral oil induced plasmocytoma) 21. As this cell line has been essential also in the hybridoma technique, its origin is important to reconstruct. While Potter expanded his myeloma tumors *in vivo* by transplantation from mouse to mouse, others had realized that myeloma cell lines growing in tissue culture would be convenient to have for many reasons, if not the avoidance of the cumbersome handling of mice. For the isolation of mutants a tissue culture cell line is absolutely essential as the precise origin of possible mutants isolated from whole mice would always remain doubtful. Luckily for Milstein and his colleagues, Horibata and Harris at the Salk Institute in San Diego had spent a lot of time trying to adapt Potter's mouse myelomas to tissue culture, an endeavour that turned out to be difficult and in most cases unsuccessful. For some reason, most likely the requirement of a compound for cell growth that is present in mice but not in tissue culture, the attempts failed. Except for MOPC 21, which grew and continued to secrete its immunoglobulin in tissue culture. Milstein obtained the tissue culture line of MOPC 21 from Harris. The cell line was cloned and one of the clones, termed P3, became the basis of all further experimentation. Milstein and his co-workers cultured the P3 cells for several weeks, isolated several thousand subclones, purified the mRNA for the heavy chains from each of them, and determined the nucleotide sequence. The results from this elaborate effort of several man-years of work were very disappointing. Only a handful of mutant clones were found, the mutations were not in the variable but in the constant region, and were mostly deletions of large sections of the gene resulting in severely truncated heavy chains that could not give rise to functional immunoglobulin molecules. Milstein and co-workers published these results as representing "the first evidence at the protein and nucleic acid levels of the existence of somatic mutations of mammalian cells". This was a wise way of phrasing, as it was clear that these mutations did not represent the putative somatic mutations that diversify antibody genes in B cells. Rather, they represented the types of rare mistakes all cells make when replicating their DNA.

The group around Mathew Scharff at Albert Einstein College of Medicine, USA, later showed that the experiment unsuccessfully attempted by Milstein and his colleagues was indeed possible. In contrast to Milstein, they had used another mouse myeloma, S107, originally generated in Melvin Cohn's laboratory at Scripps. Similar to a small number of other mouse myelomas, the antibody secreted by S107 had a fortuitous antigen binding specificity, it bound to phosphorylcholine. Other than in Milstein's experiments, Scharff and colleagues could prescreen the subclones for loss of binding to this antigen. They subcloned S107 cells on agar plates in the presence of phosporylcholine coupled to a protein carrier. All non-mutated subclones would secrete antibody that bound to the phosphorylcholine and form an opaque halo, whereas the loss mutants would fail to form this halo. Concentrating on these few loss mutants they could relatively rapidly identify and discard those that failed to secrete proper antibody molecules owing to deletion mutations such as those described by Milstein, and thus focus on mutants that secreted intact antibody molecules. In 1982 they published the amino acid sequence of the heavy and light chains of one such mutant subclone of S107 which carried a single amino acid replacement in the variable region of the heavy chain. This was the first demonstration of a spontaneous point mutation altering the amino acid sequence of an antibody molecule in the variable region and affecting its antigen binding specificity. Nevertheless, it still remained unclear if it represented the somatic diversification mechanism of normal antibody-producing B cells.

A number of other immunological laboratories did cutting edge research in post-war Europe and thus contributed to the immunological scene around Köhler, although he did not actually work there. Some of them are described in other chapters if they had an impact on Köhler's invention. In the present context I will only discuss the Department of Immunology at the Insitute of Genetics of the University of Cologne, created by Klaus Rajewsky, as it was one of the first to adopt the hybridoma technology and played a pivotal role in its improvement and dissemination.

A pioneer in the post-war development of German immunology, Klaus is the son of Boris Rajewsky, a Russian emigrant who excelled in radiation research and, as director of the then-leading institute in this field at the University of Frankfurt, was a most influential figure in German academia. He had another son, Manfred, who became a prominent scientist in German cancer research. Klaus Rajewsky studied medicine and chemistry and went as a postdoc to the Pasteur Institute in Paris to work with Pierre Grabar, inventor of immunoelectrophoresis, where he was first introduced to immunology. When he returned he joined the Institute of Genetics that was just being founded by Max Delbrück at the University of Cologne as an ambitious pilot project trying to introduce the American departmental system in a German university. Max Delbrück, famous pioneer of phage genetics of bacteria, was founding director but maintained his main laboratory at the California Institute of Technology in Pasadena, where he also continued to spend most of his time. Klaus Rajewsky then had another short but most consequential research period abroad, he joined the lab of Avrion (Av) Mitchison in London for about six months. In contrast to many other European countries, Britain had resisted German invasion and maintained a respectable academic standard through the World War II period. In Mitchison's laboratory Rajewsky acquired the technique of adoptive transfer of lympocytes in inbred mice which was instrumental in elucidating the helper function of T cells. In 1969, at age 32, Rajewsky was made full professor and head of the Department of Immunology at the Institute of Genetics. In 1971, I was fortunate to be able to join Rajewsky's department, coming from a postdoctoral position at the Rockefeller University in New York. Rajewsky's laboratory was among the first that did internationally competitive immunological research in Germany after the war. I vividly remember a day during my time at Rockefeller when I read the first paper showing that cooperation between two types of cells was required for immune responses, the first evidence for the requirement of helper cells in antibody production, published by Rajewsky and co-workers in the *Journal of Experimental Medicine*. This was the group that I had to join after returning home. In subsequent collaborative work with Mitchison, Rajewsky's group showed by the use of the adoptive transfer technique that the two cooperating cells were the B cell that secreted the antibody and the T cell helping it to do so by recognizing another part of the antigen, the so-called carrier. At the time, nobody knew what T cell help meant in molecular terms. With the studies on the helper effect of T cells Rajewsky's department advanced within a few years to an international research and training center in immunology, a status that it maintained for several decades. Rajewsky was among the first to successfully establish the hybridoma technology of Köhler amd Milstein in his laboratory. He obtained all the necessary materials and cell lines from Köhler even before the initial publication, and contributed a number of improvements to the technique over the years. Although Köhler never worked in Rajewsky's laboratory, they maintained a close relationship and much later Köhler even tried to persuade Rajewsky to accept an offer as director at the Max-Planck Institute of Immunobiology, to no avail. Rajewsky, although he received multiple attractive offers from universities and research institutes in Germany and abroad, stayed in the Genetics Institute in Cologne until his compulsary retirement at age 65, after which he moved to Harvard University in Boston.

# Chapter 3
# Köhler's entry into science

In 1994, Georges Köhler received a letter from an American journalist, Joan Oleck, 738 Union Street, Brooklyn, NY 11215, USA. Oleck indicated that she was thinking of writing a biography about Georges Köhler and requested to be sent whatever was available in terms of personal biographical information. Köhler answered by sending his CV, bibliography, and a xerox copy of a three page Science article by Nicolas Wade about the invention of the hybridome technique. His accompanying letter consisted of two sentences, one of them saying that "the enclosures are the best I am willing to provide".

The example is typical of Köhler's relationships to individuals outside of his family and perhaps a few close personal friends, if they existed. He was extremely private and notoriously reluctant to share personal feelings and private thoughts with others. This attitude is maintained like a legacy by his widow, Claudia Köhler, up to the present day. When I debated with myself whether to write this book, I had a meeting with Claudia asking her if she was willing to let herself be interviewed about her relationship to Georges as a husband and father and a private individual. Claudia refused politely but without a second of hesitation, refering to Georges' negative attitude to requests of this sort. Fabian Köhler, the youngest son, presently working at the Max-Planck Institute for Immunobiology as a postdoctoral fellow, keeps with his mother, as do the two daughters, Katharina and Lucia. Köhler as a private person will therefore transpire merely from some scattered impressions of his supervisors, colleagues, and co-workers, whom I was able to interview. Collectively the impression prevails that Köhler had close personal relationships to only a few individuals outside his family.

Georges Jean Franz Köhler was born on 17 April 1946, in Munich to a German father, Karl Köhler, and a French mother, Raymonde Köhler, nee Laporte. When Georges was 10 years of age, the family moved to Kehl, a medium size nondescript town on the German side of the German–French border near Strasbourg where Georges joined the local high school, termed

Progymnasium Kehl a. Rhein. His school performance nicely conformed to the stereotype that only poor pupils are likely to advance to outstanding scholars. The scale of marks at Progymnasium Kehl were: Excellent, good, satisfactory, sufficient, flawed, unsatisfactory. If a pupil got one of the two lowest marks in his/her report card at the end of the year, advance to the next grade was barred and the year had to be repeated. Köhler's report card at the end of his year of entry into the 6th grade of high school reads: German, sufficient; French, satisfactory; Maths, satisfactory; Biology, sufficient; Geology, satisfactory; Music, satisfactory; Arts, satisfactory; Sports, good. His marks did not significantly improve over the years. In 1964, his next to final report card at the end of the 11th grade reads: German, sufficient; French, sufficient; Maths, good; Physics, satisfactory; Biology, satisfactory; History, satisfactory; Geology, good; Music, good; Sports, excellent. Pupils also got marks for two general aspects of behavior, conduct and cooperation. While his conduct was marked as "good", his cooperation in the 11th grade received a "barely sufficient". It seems that his high school teachers were unable to raise Köhler's interest in any of the subjects, except perhaps for sports. He did only the absolutely minimum necessary to progress through high school, just barely avoiding having to repeat a grade. He did not quite make it to the level of Einstein who is said to have had to repeat a term in high school. In particular, there is no indication of any sort of special interest in biology.

He finished high school in 1965 with the baccalaureate and started in the same year to study biology at the Albert-Ludwigs University in nearby Freiburg im Breisgau. His final report card has apparently been lost but it stands to reason that his general attitude as a pupil had not drastically changed during his final term in school. What made him choose biology as a main subject is not known but there are indications that his interest in this field, as well as his general intellectual motivation, increased over the years. In addition to the regular subjects that students of biology had to take as a duty, he subscribed to voluntary activities such as higher mathematics, natural philosophy, theory of cognition in natural sciences. He also participated in a seminar series for biology students aiming to become high school teachers, a choice that he seems to have contemplated at times. However, he chose to take a diploma instead, entitling him to an academic or professional career. He performed experimental work at the Institute of Microbiology and wrote a diploma thesis on mutant strains of *Escherichia coli* unable to perform DNA repair. The experiments involved the induction of mutations by ultraviolet (UV) light, followed by isolation of mutants by single-cell cloning. While the diploma thesis did not yield any spectacular results, the subject is perhaps significant as mutations and single-cell approaches remained the center of his interest for most of his further career. During this work, he also developed a particular interest and some skills in the use of computers, and published a paper with one of his supervisors on the introduction of computer-aided teaching systems, then a controversial subject in the faculty of biology. Köhler was engaged in political student affairs, not unusual during the heyday of the students revolution in all of Europe. Nevertheless, Köhler impressed some of his professors as an unusually gifted student. One of them was Rainer Hertel, at the time chairman of the department of plant molecular biology at Freiburg University: "Köhler frequently asked intelligent questions. He was a student with whom one could have interesting discussions, on scientific and on political issues."

During his time at Freiburg University he met his wife, Claudia, nee Reintjes, born 1947 to Hans Heinrich and Katharina Margarete Reintjes in Krefeld, but raised in Wehr/Baden, a small town in a southern extension of the Black Forest bordering on Switzerland called Hotzenwald (cradle's forest). Claudia refers to her upbringing in the rough and simple woodworkers' environment of this remote region with a mixture of pride and irony. Claudia worked in Freiburg as a physician's practice assistant, a profession not requiring academic education. She is reported to have been an exceptionally pretty and charming young lady, in high demand by the young men. Georges and Claudia got married in 1968, in the 9th semester of his studies. Georges' father Karl seems to have left the family by then, his place of residence is listed as unknown in the wedding documents. Their first daughter, Katharina, was born in 1969, 13 months after the wedding and with Georges still being a student. It is reported that Georges improved his and his family's standard of living by taxi driving at night. Köhler finished his studies with a diploma (equivalent to Masters) degree in March, 1971, i.e., after 12 semesters. The regular time span for a diploma in biology was eight semesters plus one semester for the diploma thesis.

Georges got to know his PhD supervisor-to-be, Fritz Melchers, through recommendation by his professor, Rainer Hertel. Fritz is the son of Georg Melchers, prominent plant biologist and director of the Max-Planck Institute of Biology in Tübingen, dominant and charismatic figure in the Max-Planck Society and in German science in general. Fritz had obtained his PhD from the University of Cologne with experimental work performed at the Institute of Genetics, then under the directorship of Max Delbrück. He then joined Edwin (Ed) Lennox at the Scripps Research Institute in LaJolla, California, as a postdoctoral fellow. There he was introduced to immunology and was asked to study the mechanism of immunoglobulin secretion. It was already known at the time that immunoglobulin molecules existed in two forms, as surface receptors on B cells and as soluble molecules secreted by plasma cells, the final stage of B cell differentiation. Melchers and Lennox considered the possibility that secretion had something to do with glycosylation. Melchers demonstrated that, as immunoglobulins are travelling through the secretory pathway of the cell from the endoplasmic reticulum over the Golgi apparatus to the secretory vesicles and finally to the cell surface, more and more sugars were attached to them in a sequential and ordered fashion. Somewhat prematurely, he concluded from these results that stepwise glycosylation was the signal for secretion. However, while the glycosylation studies were among the first demonstrations of how proteins get glycosylated within the cell, it soon became clear that glycosylation is not a signal for secretion. Rather, non-secreted receptor type immunoglobulins are glycosylated in just the same way on their way to the cell surface. Moreover, cells whose glycosylation mechanism is poisoned by drugs, secrete immunoglobulins in just the same way as unpoisoned cells. Like so many open questions in biology, the difference between receptor and secreted immunoglobulins became clear only much later when the genes encoding them had been sequenced. Two different versions exist for the carboxy-terminal end of the heavy chain, attached to the heavy chain by alternative splicing of two final exons of the mRNA. One of these exons encodes a hydrophobic transmembrane region that fixes the molecule in the membrane, thus generating receptor immunoglobulins. The other encodes a hydrophilic sequence that is readily released from the membrane so that these immunoglobulins are secreted.

After his postdoc Melchers returned to Germany and worked as a staff scientist in the Max-Planck Institute of Molecular Genetics in Berlin. While he continued his studies in secretion, Niels Jerne had just successfully concluded his negotiations with Hoffmann-LaRoche on the foundation of the Basel Institute of Immunology in 1969 and invited Melchers to join. Melchers agreed but the Basel Institute was under construction until late 1970, and to bridge this period Melchers went to Stanford University to join the laboratory of Leonard (Len) and Leonore (Lee) Herzenberg, a scientist couple that were leading immunologists, at the time working on the first prototype version of a flow cytometer together with the company Becton-Dickinson. Flow cytometry made the detection of fluorescent signals independent of the human eye possible, and flow cytometry instruments subsequently developed to become an indispensable part of the analytical and preparative repertoires of technologies in immunology. Melchers had a good sense of foresight and wanted to get experience with the technology for later use in Basel. Graham Mitchell, an Australian scientist who had been a student with Jaques Miller at the Walter and Eliza Hall Institute in Melbourne and involved in Miller's famous experiments of the early 1960s demonstrating the role of the thymus in T cell development, was a postdoc in the neighboring laboratory of Hugh McDevitt at the same time. Together, Melchers and Mitchell used the flow cytometer trying to detect the elusive antigen receptor on T cells, unsuccessfully like many other investigators at the time.

It was in Stanford that Melchers received two letters from Rainer Hertel whom he knew from the time spent together as junior scientists at the Institute of Genetics in Cologne. Hertel, who had moved from Cologne to Freiburg as full professor, had learned that Melchers was to join the Basel Institute and offered in the first letter to sponsor Melchers' habilitation at

Freiburg University, the second thesis required by German universities to qualify for a professorial position. Hertel's interest was that after the habilitation Melchers would become an external member of the faculty of biology with a teaching obligation in Freiburg University, that did not have an official teaching program in immunology in its repertoire, although students were sporadically exposed to immunological subjects. In the second letter, Hertel recommended to Melchers a young student, Georges Köhler, who wanted to do a PhD thesis in immunology, and who had impressed him, Hertel, as an intelligent and promising candidate. Nevertheless, Hertel retrospectively added: "I was somewhat reluctant with this recommendation. Köhler being rather introverted and quiet, I could not be absolutly sure if he had the brilliance that seemed to be a prerequisite for being successful in the Basel Institute. But I did not tell this to Fritz." Melchers moved to Basel in 1971, and accepted Köhler as his first PhD student, based on a positive impression gained during an interview and, perhaps more importantly, on the recommendation of Hertel.

Melchers gave Köhler two choices of topic for his PhD thesis. Melchers' preferred choice was a project within the long-standing and established program of his laboratory on immunoglobulin secretion and the role of glycosylation in this process, with a possible extension to experiments on immunoglobulin synthesis in resting and activated B cells. Köhler apparently had rather firm preferences, he said he was not interested at all in secretion and flatly refused the project. He took the second choice, a project that was aimed at estimating the heterogeneity of antibodies made against a single epitope of a protein antigen. The antigen in question was β-galactosidase, an enzyme of the bacterium *Escherichia coli* that catalyzes the fermentation of galactose providing the bacteria with energy. Because the gene (gal) encoding this enzyme can be transferred by lambda phage from one bacterium to another, the enzyme had been for a long time the center of interest of bacterial geneticists and was well known to Melchers from the times of his thesis at the Genetics Institute in Cologne. Many mutants of gal have been isolated, including those which encode enzymes with reduced catalytic activity. Antisera have been raised against the wildtype and mutants for reasons of identification. Intriguingly, in the presence of antibodies to the wildtype enzyme some of the mutant enzymes showed increased enzymatic activity, in some cases approaching that of the wildtype. Melchers explaines this as follows: "*The mutant enzyme oscillates between a frequent mutant- (enzymatically inactive) and an infrequent wildtype- (enzymatically active) protein conformation. Antibodies specific for the wildtype conformation fix the wildtype form, like a stapler, preventing the reversion of the active to the inactive conformation. In other words, a mutant enzyme molecule which is attached to an antibody molecule recognizing the active site of the wild-type enzyme gets 'frozen' in the active conformation of the wild-type enzyme.*" Melchers had started to work on such antibodies at the MPI in Berlin together with Walter Messer, and it may have been the aspect of antibody specificity and diversity that made Köhler prefer this project. Since antibodies of this type presumably recognize very restricted epitopes associated with the enzyme's active site, Melchers and Köhler reasoned that they were directed against a single protein epitope. The question that Köhler was supposed to answer was how many different antibodies of this type an animal can make. He thus immunized several species of laboratory animals, including rabbits, mice, and rats, with β-galactosidase, collected their sera at various times after immunization and analyzed the diversity af antibodies they had made. The experimental approach to antibody diversity in vogue at the time was isoelectric focussing, in which complex mixtures of antibodies are subjected to an electric field in a gel slab in which a pH gradient is established by a matrix of ions. Antibodies of different structure differ in their net charge and, consequently, in their isoelectric point, i.e., the pH at which the positive and negative charges are equal. At its isoelectric point the antibody acquires a zero net charge and therefore stops migrating in the electric field. The typical result for antibodies even against single well-defined antigens is a gel slab with large numbers of bands that usually are far too complex to provide meaningful information. The problem Köhler had to master was how to focus the analysis on antibodies that restored the activity of inactive mutant enzymes. The experimental system had obviously been chosen because it offered a straightforward approach to achieve this goal by exploring the ability of such antibodies to induce mutant enzymes to acquire activity. After isoelectric focussing, the gel slabs were bathed in a solution containing the mutant enzyme, allowing the enzyme to bind to the antibodies in the gel. Thereafter, the gel slabs received a second bath in a solution containing an artificial substrate that gives a color reaction when catalyzed by the active enzyme. In this way, bands in the gel that corresponded to antibodies able to induce mutant enzyme to acquire activity showed up in green color, whereas bands of antibodies that merely bound the enzyme without inducing activity remained without color. Köhler developed this demanding technique to great resolution and reliability. Melchers: "Georges certainly was a deft and skillful worker in the lab, a punctilious experimentalist."

The results Köhler obtained varied from species to species and from mutant enzyme to mutant enzyme. Three different enzyme mutants were studied, with differing results indicating that restoration of activity for each mutant was achieved by antibodies against a different epitope. More pronounced were the differences between rabbits and mice. While individual mice produced large arrays of different antibodies, each individual rabbit responded with only a few clones which, although remaining constant over long periods of time, varied from rabbit to rabbit. Niels Jerne did not seem to be very impressed with Köhler's results, presumably because he did not expect anything other than a pronounced heterogeneity of antibodies even against a defined epitope. Others in Basel were interested, most of all Steven Fazekas de St. Groth, an Australian immunologist who was famous for first observing the phenomenon of "original antigenic sin". This term describes the fact that upon multiple annual infections with influenza virus, antibodies in any given year always include a strong fraction that cross-reacts with last year's virus but no antibodies that cross-react with the virus of the following year. The explanation is that the ongoing infection activates cross-reacting memory B cells from the previous year more readily than naïve B cells expressing antibodies to the actual virus that do not cross-react with the previous virus or cross-react with next year's virus. Like many immunologists of that time Fazekas de St. Groth was fascinated with the endless complexity of the antibody problem, and Köhler's results were another intriguing example. Köhler tried mathematics to estimate antibody diversity in his experiments. With the help of Fazekas de St. Groth and Charles Steinberg, intellectual guru and leading theoretician at the Basel Institute, they employed Poisson statistics based on the repeat frequencies by which certain antibody clones could be found in different individuals. In public literature the results of Köhler's thesis are often summarized by statements such as "a single individual can make a thousand different antibodies to a single epitope". This is certainly a gross oversimplification but Köhler himself used it in lectures to general audiences. The method of analysis was not sufficiently precise to permit calculations of such accurate figures, and the statement does not do justice to the results on rabbits that only seem to produce a few antibodies per individual. Melchers today: "I still do not understand these differences between species."

While Köhler and Melchers seem to have had a good understanding in scientific matters during Köhler's thesis, there were certain facets of Köhlers's approach to life in general and work in particular that Melchers was not at all happy with. Most of all, he objected to Köhler's relaxed and lazy working attitude. It usually goes without saying that a young student or postdoc works considerable overtime. Not only does this demonstrate motivation and excitement about the work, for most of the time it is simply a matter of necessity as many experimental protocols do not fit into an 8-hour day or a 5-day week schedule. Not so Köhler. Rarely did he work at night, not to speak of coming in at weekends. On the contrary, Köhler seemed to enjoy family life and sired his second and third child during the time of his thesis: Lucia, born 7 January 1972, and Fabian, born 22 December 1972. After each new child, Köhler went to see Melchers to ask for a rise in salary, which Melchers granted but with considerable toothache. Melchers: "Somehow he seemed to take things in the wrong order." When it came to publication of the thesis work, Melchers asked Köhler to publish his results without him, Melchers, as a co-author. This good old academic tradition, now outdated, requires that the student mentions the name of his doctorfather and thanks him profusely for encouragement and support in the acknowledgement section of the paper. Instead, Köhler thanked both Melchers and Milstein (the paper was written while Köhler was in Cambridge) for helpful discussions. To make it worse, he thanked the technician of the laboratory, Dorothee Jablonski, for technical assistance and "moral support". Melchers remembers another irritating moment that took place during Köhlers thesis defense examination at Freiburg University in April, 1974. Towards the end of the exam, that had so far gone well but not outstanding, the chairman of the committee asked Köhler if he was willing to answer another more demanding question. A correct answer would be necessary to obtain the top mark, *summa cum laude*. Köhler answered that he would rather not be asked that question and was happy with the result as it were. Köhler thus obtained his doctoral degree with the mark, *magna cum laude*, which is the second in the scale and given to the vast majority of doctoral candidates. After the exam was finished, Melchers had to endure suggestions from the chairman, the rather traditionally minded professor of cell biology in Freiburg, Peter Sitte, how to educate future PhD students towards a more ambitious attitude. Although Melchers retrospectively prefers to emphasize the comical aspects of these events, one can imagine that on some occasions the relationship between Köhler and Melchers had to endure considerable stress.

While Köhler's thesis added to the growing number of examples of the astonishing hetrogeneity of antibody responses, it contributed little to the understanding of this phenomenon. What were the genetic factors generating and controlling antibody heterogeneity? What made different animals, even genetically identical animals such as inbred mice, respond to the same antigen with different, individually specific, arrays of antibodies? While in the 1970s the debate among followers of the germ line and the somatic mutation theories was still ongoing and far from being settled, scientists at the Basel Institute were under the influence of Jerne and thus mostly thought and argued in favor of somatic mutation theories. As a result, in their discussions on how to continue their studies, Köhler and Melchers frequently discussed the famous "fluctuation analysis" of Max Delbrück and Salvador Luria as an example of how one should search for mutations in antibody genes. Fluctuation analysis is a means to identify spontaneous mutations in bacteria. A clone of *E. coli* bacteria, sensitive to the antibiotic penicillin, is divided up into many subcultures of small numbers of bacteria. These are further grown for several generations and then treated with penicillin. While most bacteria die in the antibiotic, some penicillin-resistant mutants have arisen that continue to grow. The typical result is that the number of penicillin-resistant mutant bacteria strongly varies between subcultures, it fluctuates. This means that the mutations leading to penicillin resistance have occurred in individual bacteria in each subculture at different time points prior to the addition of penicillin, suggesting that mutation is a random and spontaneous process. Moreover, it permitted the estimation of mutation frequencies of bacterial genes. Melchers and Köhler thought that experiments like fluctuation analyses with antibody-producing cells was a way to approach the genetic diversification mechanism of antibody genes. Still in Basel, Köhler tried to visualize individual cells secreting antibody to β-galactosidase from the spleens of immunized mice by placing them in agar gel soaked with mutant β-galactosidase, with the aim of making them visible by a color halo caused by the secreted antibody activating the enzyme in the gel. The assay did not work, it was too insensitive. Moreover, even if the assay had worked, the problem remained that normal B cells did not divide outside the body. Hence, above all other technical problems, the project required clones of B cells that secreted specific antibodies and that could be grown in culture. There were two laboratories in the world that were close to establishing such B cell cultures, that of Mathew Scharff at Albert Einstein College in New York, and that of Cesar Milstein in Cambridge. Scharff's laboratory was further advanced because they could culture and clone a mouse myeloma line, S167, secreting antibody specific to phosphorylcholine. Milstein could culture another mouse myeloma, MOPC 21, which secreted antibody without a known specificity. In spite of Scharff's advantage, Köhler chose Milstein. He did not like the idea of living in the US.

# Chapter 4

# The quest for monoclonal antibodies

Long before Köhler finished his thesis, immunologists had been keen on obtaining monoclonal antibodies. There were a number of reasons, not only to understand genetic diversification. Despite the vast heterogeneity of normal immunoglobulins in serum, remarkable progress had been achieved in elucidating the main structural features of antibody molecules. For example, in the late 1950s and early 1960s, Rodney Porter's and Gerald Edelmann's experiments had revealed the symmetrical 4-chain structure, consisting of two light and two heavy chains. The molecular weight of light chains corresponded to a polypeptide chain of about 220 amino acids, that of heavy chains to about twice that length. This basic structure was common to all antibody molecules, independent of class or antigen specificity. Nevertheless, the pressing question remained how antibodies discriminated among the millions of antigens to which they could specifically attach. To explain this phenomenon, the most likely assumption was that antibodies to different antigens differed in their amino acid sequence. How pronounced were these differences? Were these putative differences located in the heavy chains or in the light chains, or perhaps in both? At what positions in the amino acid sequence did these chains differ from one another? To determine the amino acid sequence of a protein was a major and long-term endeavor at the time. The analysis required grams of material and was done by cleaving off and determining amino acids one by one, from that end of the polypeptide chain now, but not then known to be the variable region. Attempts to obtain amino acid sequence information on normal immunoglobulins isolated from serum usually ceased to provide interpretable results after a few amino acid positions into the sequence, as they revealed multiple different amino acids at each of these positions in both heavy and light chains. This type of result did not allow for the reconstruction of the amino acid sequences of individual antibody polypeptide chains in the mixture and thus once again only reinforced the concept of the vast molecular heterogeneity of antibodies. Meaningful sequence information was to be expected only from homogeneous antibody preparations.

But how could one get such antibodies? A handful of laboratories engaged in this race, choosing different avenues, not knowing which one would lead to success. Henry Kunkel, a genial scientist/physician at the Rockefeller University in New York, was among the first to realize that myeloma proteins resembled normal immunoglobulins. Moreover, he and his collaborators presented evidence that myeloma proteins were homogeneous in structure and thus likely to stem from a single clone of plasma cells, i.e., were of monoclonal origin. The Rockefeller University supported a small hospital with about 25 beds that scientists with an MD degree could use to study human diseases. Patient care at the hospital was excellent but free of charge for the patients, and patients admitted to the hospital could therefore be strictly selected to fit the criteria of the various projects under study. Kunkel studied patients with multiple myeloma, a malignant tumor growing mostly in the bone and the bone marrow and associated with high concentrations of a novel type of protein in the serum. Upon electrophoretic analysis these proteins migrated in the region of immunoglobulins but did not show the electrophoretic heterogeneity typical of immunoglobulins. Instead, these proteins showed electrophoretic homogeneity and each patient's protein had a different characteristic electrophoretic mobility. These proteins were referred to as paraproteins because their relationship to immunoglobulins was suspected but not certain.

A main tool in the study on the relationships between myeloma proteins and normal immunoglobulins were anti-idiotypic antisera. At the time, the determination of an unknown type of protein was not at all a straightforward task. Amino acid sequence analysis was in its infancy and DNA sequencing did not exist. One of the main techniques to identify an unknown substance therefore was to test its reactivity with antisera prepared to known molecules. If the unknown substance reacted with an antiserum of known specificity, it was likely to have structural similarity to the substance recognized by the antiserum. The alternative was to produce antisera to the novel substance by injecting it, for example, into a rabbit. The antiserum of the rabbit could then be tested for reactivity with molecules of known composition. Kunkel and his collaborators performed such experiments with paraproteins isolated from the sera of a large number of patients. It soon became clear that paraproteins reacted with antisera prepared against normal immunoglobulins, suggesting structural relationships to immunoglobulins. More interesting was the reverse experiment: antisera prepared to paraproteins also reacted with normal immunoglobulins, but a fraction of the antibodies in each antiserum was specific to the paraprotein and did not bind to normal immunoglobulins. Such antibodies recognized what Kunkel dubbed the "individual antigenic specificity" of the paraprotein, another term but meaning the same as the idiotype, i.e., the unique structure of the antigen attachment site of the paraprotein. In normal heterogeneous immunoglobulins, antibody molecules having the same antigen attachment site as the paraprotein were either totally absent or so rare that they could not be detected. The presence of an idiotype was thus taken as evidence that the paraprotein was a monoclonal antibody with all molecules bearing the same antigen attachment site. However, the drawback of all of these studies was that paraproteins, though being acceptable as examples of monoclonal antibodies, lacked a known antigen specificity due to their origin from random tumors rather than from deliberate immunization.

Experiments such as those performed by the Kunkel laboratory, fascinating as they were to many immunologists, fell short of satisfying hard core molecular biologists and biochemists who were keen on understanding the precise structure of the antigen attachment site of antibodies. Gerald Edelman, also working at the Rockefeller University, belonged to this group. He had previously elucidated the 4-chain structure of antibodies by splitting the disulfide links that hold heavy and light chains together using chemical reduction and alkylation reactions. The present aim of Edelman's laboratory was to determine the amino acid sequence of antibody polypeptide chains, and they had already begun the herculic task to sequence an entire myeloma protein. Edelman only believed in hard structural data and considered results obtained with idiotypic antisera as entirely inconclusive. At a formal lecture given to Rockefeller scientists in the late 1960s, Edelman publicly said that he could see no merit in experiments involving the production of antibodies against antibodies, thus expressing a certain degree of contempt for the work of the Kunkel laboratory. A consequence of the lack of friendship betwen Kunkel and Edelman was that Kunkel, whose laboratory had not established amino acid sequence analyses in the early 1960s, collaborated with a young German postdoctoral fellow, Norbert Hilschmann, working with Lyman Craig in another Rockefeller laboratory equipped to perform amino acid sequence analysis. Hilschmann obtained from Kunkel a pair of paraproteins that consisted only of light chains, so-called Bence-Jones proteins, and that were to become the first fully sequenced antibody polypeptide chains. As described in Chapter 1, the amino acid sequence analysis of this pair of light chains, published in 1965, led to the pathbreaking discovery that immunoglobulin light chains have variable and constant regions. Edelmann's sequence of an entire immunoglobulin molecule was published in pieces in several papers between 1968 and 1970. Not only did it take much longer, it also had less impact as it lacked the clue of Hilschmann's experiment allowing the comparison of two different light chains.

The question is often asked why Hilschmann was not included when the Swedish Academy of Sciences decided to award the 1972 Nobel Prize for Physiology or Medicine to Porter and Edelmann for their work on antibody structure. There is no doubt whatsoever that Hilschmann's discovery of V- and C-regions was of similar outstanding importance as was the the 4-chain structure in the elucidation of antibody structure-function relationships.

Moreover, Hilschmann's experiment ranked among the few that immediately generated a new way of thinking, a new paradigm. This must have been clear to the Nobel committee in 1972, seven years after the publication. It is possible that the Nobel committee did not find a solution for Craig, who had never worked on antibodies before or after. Alternatively, it is often speculated that the reason was a faux pas that Hilschmann is reported to have committed at a conference in 1964 where all important researchers in the antibody field were present. When Hilschmann presented his still unpublished and incomplete sequence results in his talk, many in the audience were keen to copy down the partial sequences for comparison with their own less advanced sequence results on other antibody preparation. According to hearsay of participants at this meeting, Hilschmann ran his data slides very fast so that copying was not possible, and did not slow down when asked to do so. For many years thereafter, Hilschmann's behavior was not forgiven by the scientific community in the US. Soon after his paper was published, he returned to Germany and in 1971 became the director of the Max-Planck Institute for Experimental Medicine in Göttingen. He continued to do cutting edge work but often spitefully published this in somewhat obscure scientific journals. For example, in the early 1980s Hilschmann's group was again the first to have determined the amino acid sequence of a major histocompatibility class II polypeptide chain, and published this in German language in *Hoppe Seyler's Zeitschrift für Physiologische Chemie.*

A few human myeloma proteins were fortuitously found to bind specifically to antigens. One of those was a so-called rheumatoid factor, i.e., it resembled the autoantibodies against human immunoglobulins frequently found in patients with rheumatoid arthritis. Being a human immunoglobulin itself, this myeloma protein reacted with itself under certain conditions forming a precipitate and lending itself to crystallization. Using this myeloma protein, David Davies at the National Institute of Health in Bethesda was the first to analyze crystals of an antibody molecule. Analysis was done by electron microscopy rather than by X-ray diffraction, because the crystals were not of sufficient quality for the latter, more precise technique. Nevertheless, David Davies' images of the antibody molecule were a striking confirmation of the predicted three-arm structure, even though it looked more like a T than a Y, the angle between the two antigen binding arms approaching 180°.

Even though David Davies' electron micrographs confirmed once again that the antigen attachment sites were located on the two arms of the antibody that were formed by the paired heavy and light chains, the few examples of human myeloma proteins with known antigen binding specificity were not sufficient for a systematic analysis of the structure of the attachment site and the amino acid sequence variations suspected to account for antigen specificity. However, the collection of myeloma proteins with known antigen binding specificity was significantly enlarged when Michael

Potter at the National Institute of Health found that myeloma tumors could be deliberately induced in inbred mice by injection of mineral oil into the peritoneal cavity. The tumors were induced in the tissues associated with the intestines and, consequently, most of them secreted homogeneous immunoglobulins of the class characteristic for mucosal secretion, IgA. The tumors formed copious amounts of ascites fluid and could be transplanted from mouse to mouse so that there was no limitation of the amounts of protein available. While the tumors still arose at random and antigen specificity was not deliberately inducible, a fair proportion secreted myeloma proteins that bound to bacterial products, mostly carbohydrates. The high proportion of myeloma tumors with such specificities was suspected to result from an enrichment in the gut tissues of plasma cells with specificity for components of the intestinal flora. In addition to bacterial components a number of these mouse myeloma proteins reacted with the small chemical compound dinitrophenyl, one of the chemicals originally used by Landsteiner to reveal the unlimited versatility of antibodies.

Other laboratories engaged in looking for monoclonal antibodies by using special immunization procedures. Richard Krause at the Rockefeller University had received his scientific training in the laboratory of Maclyn McCarty and Rebecca Lancefield who were engaged in studying *Streptococcus* and the infections it caused. McCarty had been a member of the laboratory of Oswald Avery who was interested in pneumococcal infections and in the course of elucidating the principles of bacterial transformation discovered DNA as the substance responsible for inheritance. Krause took a closer look at the antibodies to streptococcal isolates that the laboratory routinely produced in rabbits in order to determine the groups and types of streptococci isolated from patients. By doing electrophoretic analyses of such rabbit antisera, Krause discovered that some of them looked just like the sera from patients with multiple myeloma, and his laboratory subsequently demonstrated the molecular uniformity of these antibodies. I was a postdoctoral fellow in Krause's laboratory at the time when it was decided to give full capacity to the study of homogeneous antibodies to streptococci in rabbits. It was found that the antibodies were directed to the carbohydrate moiety of the streptococcal cell wall. Most importantly, as such antibodies were not only homogeneous but also produced in large quantities, it was possible to obtain amino acid sequences, in this case antibodies deliberately induced to a well-characterized antigen. Thomas Kindt and Dietmar Braun, two other postdoctoral fellows in Krause's laboratory at the time, spent much of their further careers with sequencing of streptococcal antibodies, the latter in the Basel Institute of Immunology overlapping in time with the young Georges Köhler working there as a PhD student. Induced antibodies of molecular homogeneity were described by a few other laboratories, but the parameters that had to be fulfilled in order to produce homogeneous antibodies by deliberate immunization were never fully understood. Antigens that under certain conditions have given

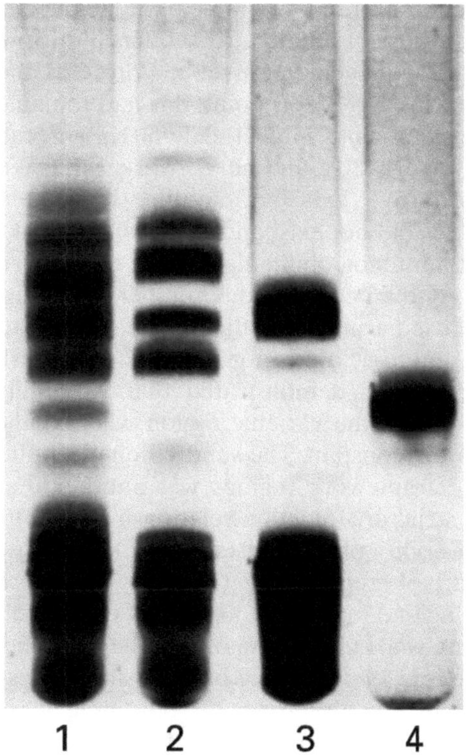


Figure 8
Polyacrylamide electrophoresis of different immunoglobulins to assess light chain heterogeneity. 1: normal rabbit immunoglobulin, 2: heterogeneous rabbit antibody to *Streptococcus*, 3: homogeneous rabbit antibody to *Streptococcus*, 4: human Bence-Jones protein. The immunoglobulins were reduced to open up the disulfide bonds and to separate heavy and light chains. They were then subjected to electrophoresis in polyacrylamide columns (bottom to top), separating proteins according to size and charge. The light chains migrate to the middle of the column. Normal immunoglobulin and the heterogeneous antibody show several bands of light chains. In contrast, light chains of the homogeneous antibody and the Bence-Jones protein migrate as a single band. The heavy chains migrate only a short distance and are therefore not well separated at the bottom part of the columns. The Bence-Jones protein contains no heavy chain (from Eichmann et al, 1970).

rise to homogeneous antibodies included compounds as different as streptococcal and pneumococcal carbohydrates, myoglobin, angiotensin, phosphorylcholine, and synthetic chemicals such as azabenzoate and dinitrophenyl. A common denominator among those antigens was obviously missing. The importance of genetic polymorphisms was revealed by observations in mice suggesting that some but not all inbred strains may produce homogeneous antibodies to antigens such as streptococcal carbohydrate or dextran, but the nature of the genetic control appeared to be multigenic and not readily unravelled.

While the search for a reliable source of inducible monoclonal antibodies was going on, a host of amino acid sequence information had already accumulated by the sequencing of myeloma proteins and a few antibodies in various laboratories. Up to this point, this had remained scattered information without significant impact. Elvin Kabat, then at New York University in New York City, and his co-worker T.T. Wu were the first to realize the informative value of systematic sequence comparisons. Electronic databases did not exist and so Wu and Kabat collected all available sequences of immunoglobulin light chain variable regions and aligned them by hand. They then counted the number of different amino acids at each position and used a simple mathematical formula to assign a variability index to each amino acid position. The result, published in 1970 as the so-called Wu-Kabat plot, demonstrated that variability accumulated in three short segments of the variable region whereas the intervening segments were relatively constant. The subdivision of antibody V regions into hypervariable and framework regions was another quantum leap in the understanding of structure–function relationships of antibodies. Not much later a similar variability plot was established for heavy chains, also showing three hypervariable regions separated by framework segments. Kabat continued collecting and publishing hard copy compendia of new antibody sequences for many years until the databases began to become available on the internet in the mid 1980s. On the basis of similarities among hypervariable regions of otherwise different antibody molecules Kabat developed the minigene hypothesis that postulated that hypervariable regions were encoded by short gene segments that would be inserted into longer genes encoding the framework regions. Like so many theories in immunology the minigene theory turned out to be wrong. In contrast, the concept of hypervariable and framework regions was unequivocally confirmed by X-ray diffraction studies of crystallized antibodies and antigen–antibody complexes. Such studies became possible only much later when monoclonal antibodies could be produced to any chosen antigen according to the hybridoma technology of Köhler and Milstein. In the folded immunoglobulin molecule the six hypervariable regions of both light and heavy chains come to lie in close proximity and together form the continuous hypervariable surface of the antigen attachment site at the ends of the antigen binding arms.

The amino acid sequence studies on the various sources of monoclonal antibodies had brought significant progress in understanding the nature of the antigen attachment site and its ability for discrimination of myriads of antigens. The six hypervariable regions of heavy and light chains together comprised a minimum of some 40 hypervariable amino acid positions. If each of these positions could be occupied by the 20 different amino acids that exist in nature, the number of different combining sites would be $20^{40}$, an unimaginably large number. In reality this number would be somewhat smaller as there are structural constraints that would not allow all amino acids at all positions. Nevertheless, even if only 10 different amino acids

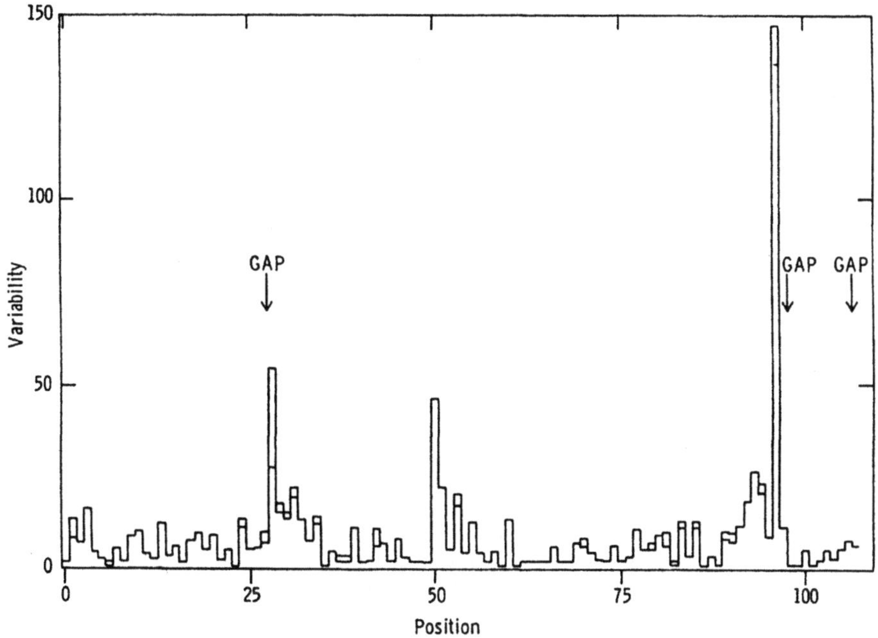


Figure 9
Amino acid sequence variability for immunoglobulin light chain variable regions calculated from all sequences known by 1970. The variability (vertical axis) is calculated for each amino acid position (horizontal axis) as the ratio of the number of different amino acids observed over the frequency of the most frequent amino acid. Note the three hypervariable regions as opposed to the limited variability of the framework regions. GAP: positions at which an amino acid may be missing (from Wu and Kabat, 1970).

were allowed at each hypervariable position the number of possible attachment sites would be a 1 with 40 zeros, more than enough to cover the universe of antigens.

Studies on antibodies at the protein level, in spite of years of effort, had little to contribute to the problem of how antibody diversity was generated genetically. Were all antibody V-genes present in the genome or were they diversified by somatic mechanisms that would alter antibody V-genes randomly and individually in each antibody-producing B cell? An important conclusion was that V region amino acid sequences could be divided into subgroups that were sufficiently different from one another to exclude their origin by somatic diversification of a single gene. The number of V-region subgroups, more than a dozen for heavy and for light chains each, therefore seemed to reflect the minimum number of inheritable V genes present in the genome. Upon comparisons of V-region sequences that belonged to the same subgroup, the differences observed could be interpreted as somatic mutations. However, the results could also be explained by differ-

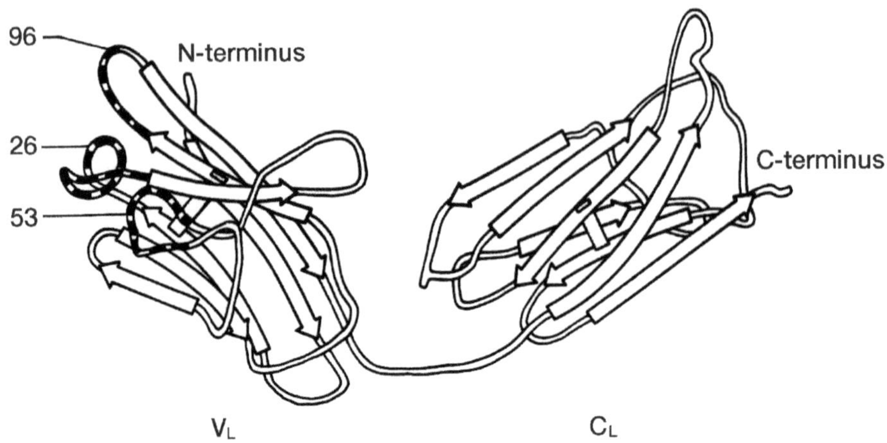


Figure 10
Three-dimensional structure of an immunoglobulin light chain as deduced form X-ray analyses of a crystallized antigen-binding papain fragment. The $V_L$ and $C_L$ regions form separate domains. A domain consists of two stacked sheets (β-sheets) of 3–4 antiparallel strands each, the two sheets being connected by a disulfide bond. The $V_L$ domain differs from the $C_L$ domain by the three hypervariable regions that coincide with three loops sticking out of the free end of the $V_L$ domain. Together with three similar hypervariable loops of the heavy chain V region, not shown in this figure for the sake of clarity, they form the antigen attachment site (from Capra and Edmundson, 1977).

ent genomic V-genes and remained therefore inconclusive. No less inconclusive was the information that came from idiotypic studies on induced antibodies in mice, in which I was temporarily involved. Owing to the work of Jaques Oudin at the Pasteur Institute in Paris it was already clear that not only myeloma proteins but also induced rabbit antibodies could bear idiotypes. Such idiotypes were specific for the antibodies of one rabbit and were not shared by the antibodies to the same antigen in other rabbits. Now it was found that certain antigens, such as streptococcal carbohydrate, phosphorylcholine, dextran, or arsonate, induce antibodies in certain inbred strains that carried a predictable idiotypic marker. The predictable idiotype was borne by the antibodies of all individual mice of one inbred strain but not by that of most other inbred strains. In mouse breeding experiments, such idiotypic markers segregated as Mendelian alleles and in some cases could be mapped relative to one another. Originally believed to represent individual V-genes, it was later realized that for antibodies to share idiotypic markers it was sufficient to share similar but not necessarily identical V-regions. Nevertheless, in some instances such idiotypes were useful as V-gene markers and these studies also suggested that the number of different inherited V-genes was considerable. However, these studies also failed to distinguish between germline and somatic theories of antibody diversity.

Experiments using idiotypes as V-gene markers were simultaneously performed by a number of laboratories during the initial years of the 1970s. As they failed to provide unequivocal answers to the burning questions in immunology, the time was rich in argumentation and theory making. A highlight was a conference organized in 1976 by Niels Jerne, then Director of the Basel Institute of Immunology, and Hilary Koprowsky, then Director of the Wistar Institute in Philadelphia, in a hotel with a three star restaurant, Ousteau de Beaumaniere, in Les-Beaux-de-Provence, south of France, which I had the privilege to attend as one of about 15 participants. The outstanding cuisine and the almost unreal pleasant surroundings fostered two delightful days of an unrepeatable intellectual exchange. The proceedings were later published in the form of a verbatim transcript by the Basel Insitute of Immunology, entitled "Idiotypes, What They Said at the Time". In the foreword, Niels Jerne wrote: "*It shows how biologists, confronted with poorly defined concepts and baffling findings, try to grope their way towards clarification. Some ideas are dismissed as phantasy, some experiments are discounted as unsound. Some reagents are unreliable, some generalizations as unadmissable – and still the debate is pervaded by a general feeling that all this is important...*" Today, with scientific technology providing the investigator with mostly unequivocal answers, ingenuity and conceptional phantasy are mostly logistic and technical, concerning the fastest and least elaborate way to a solution of a scientific problem. While the ultimate goal of science, the finding of the truth, is in clearer focus today, some of us certainly miss the argumentative attitute of the old times.

Clearly however, many investigators were deeply dissatisfied with the speculative approach to the antibody problem. As a result, there was a growing realization that in order to understand antibody genetics it was necessary to investigate antibody genes rather than antibody proteins. The initial attempts in this direction utilized RNA probes for individual light chain variable regions, isolated from myeloma cells, and employed a technique called hybridization kinetics. The RNA probe is radiolabelled and mixed with total genomic DNA that has been heated to separate the two strands in order to permit hybridization of the RNA probe to complementary V-genes. The greater the number of complementary V-genes the RNA probe finds in the DNA for hybridization, the faster the kinetics of hybridization. Essentially all the experiments expertly performed with this technique suggested that a V-gene probe hybridizes only to a single gene in the DNA. This was unexpected as already the number of V region subgroups had suggested at least several dozens of V-genes in the genome. It was then argued that a V-gene probe presumably hybridized poorly to V-genes belonging to different subgroups, because there are too many nucleotide sequence differences between subgroups which prevent efficient hybridization. However, this argument was speculative and not readily supportable by solid experimentation. The data therefore eventually supported the concept that the genome contained a single V-gene per V region subgroup, but did not take the matter much further.

So indirect experiments even using nucleic acids did not seem to help either. There remained two direct ways to solve the antibody question. One involved simple brute force and consisted of the systematic nucleotide sequencing of the immunoglobulin loci in genomic DNA. The magnitude of this endeavor is hard to convey. It requires dozens of co-workers over many years with few exciting results to be expected, just millions of nucleotides in a row. Only a few investigators had the resources and patience to tackle this task including, among others, Leroy Hood at California Institute of Technology in Pasadena, Susumo Tonegawa at the Basel Insitute of Immunology, Phil Leder at the Massachusetts Institute of Technology in Cambridge, USA, David Baltimore at Harvard University, and Hans Zachau at the University of Munich. Owing to their painstaking work over many years, the genetic composition of the immunoglobulin loci has been elucidated, much before this information became available in the context of the human and mouse genome projects. The pivotal result from these studies was that the inherited V-gene segments are insufficient to encode the observed phenotypic diversity of antibodies. Many antibody V region sequences could not be accounted for by V-genes present in the genome. Thus, in addition to somatic recombination of gene segments, there should be somatic hypermutation. The second approach was to observe somatic antibody gene diversification directly. This required monoclonal antibody producing B cells that mutated their antibody V genes, and the race for the optimal type of monoclonal B cells was soon in progress. Normal B cells that can be isolated from blood or lymphoid organs can be kept alive only for a few days in tissue culture. In addition, normal B cells are heterogeneous, every single one has different antibody genes. Some myeloma cell culture lines existed but the immunoglobulins they produced seemed stable, i.e., they seemed to have lost the ability to somatically diversify their antibody genes. Another source of B cell tumors could be obtained by transformation with Abelson virus. The cell lines obtained were immature in nature and did not secrete antibodies, nor could they be induced by immunization. No promising source of B cells was availabe for the study of somatic antibody diversification. The race for monoclonal B cells was on. Who would be the winner?

# Chapter 5
# Cell fusion

In 1960 Georges Barski, Serge Sorieul, and Francine Cornefert, working at the Institut Gustave Roussy, Laboratoire de Culture de Tissus et de Virologie, Villejuif, France, observed for the first time that cell fusion occurred when two different tumor cell lines, derived from two different inbred strains of mice, were grown as a cell mixture in tissue culture. Barski was interested in the vastly different morphological properties of tumor cell lines including their abnormal chromosomes. He performed coculture experiments to study whether "transfer of characters" can occur from one tumor cell to another. Sorieul was a Masters student in the laboratory and had to perform karyotype analysis, a method by which one can visualize and identify the chromosomes present in a cell. He observed the occurrence of a new type of cell in the mixed cultures, exhibiting the karyotypic properties of a hybrid, namely the presence of marker chromosomes of the two parental cell lines as well as nearly twice the number of chromosomes. The observation was essentially fortuitous, cell fusion was not intended, nor were any fusion-inducing reagents used or even known at the time. The results were published in French but nevertheless raised strong international attention. Sorieul then joined Boris Ephrussi at Western Reserve University in Cleveland, Ohio, to work for a PhD. There he repeated the original finding with additional tumor lines, demonstrating that cell fusion was a rare but nevertheless predictable event in tissue culture. The findings were published in several papers between 1960 and 1962, convincingly documenting cell fusion by karyotype analysis. The fused cell lines contained most of the chromosomes of both parental cells, they were "heterokaryons". The findings were soon repeated by others. The first one was the famous cell biologist Leo Sachs at the Weizmann Institute of Science in Israel. He published a paper in *Nature* in 1963, in which he and his co-worker David Gershon described the fusion of two mouse tumor lines derived from strains that differed in their histocompatibility types. Successful fusion was demonstrated by showing that the fused tumor cells expressed both parental histocompatibility antigens. Whereas the two parental tumors could be grown in mice of the homologous strain but were rejected by mice of the heterologous strain, the fusion-derived tumor was rejected by both parental strains but could be grown in F1 mice.

Why would anybody be excited about fused tumor cells? The reason was that mammalian cell fusion seemed to offer the possibility of genetic analyses in mammals without the need for time-consuming breeding experiments, not to speak of humans that cannot be bred experimentally. The approach was called somatic cell genetics and was important in the times before genome analysis, but restricted to primitive organisms. Existing model organisms in somatic cell genetics were fungi such as *Penicillium* or *Aspergillus*, or yeast. These organisms can multiply asexually by cell division but can also mate by cell fusion, depending on environmental or culture conditions, the so-called "parasexual cycle". Similar to sexual mating in higher organisms, the chromosomes and genes of the mated somatic cells recombine and segregate again, thus giving rise to novel genetic patterns in the progeny. Somatic cell mating was used to generate gene maps by recombination or complementation analyses. By recombination analysis one could estimate the distance between two genes on a chromosome, by complementation analysis one could determine if a newly observed mutation had occurred in a novel gene or in genes already known. Similar analyses were also used in bacterial phage genetics and were among the most successful means to determine mutant genes, leading to surprisingly detailed genomic maps of the organisms lending themselves to such studies. The observation of spontaneous cell fusion in mammalian cells resembled the parasexual cycle of fungi and stimulated phantasies that cell fusion could be used to perform recombination and complementation analyses in mammals in which genetic analyses so far had to rely on sexual reproduction. In addition, it was noted that many fused mammalian cells tended to lose some of their chromosomes, providing a further means to localize genes on particular chromosomes. Most importantly, it suddenly seemed possible to study mutated genes responsible for heritable human diseases. In the original experiments, mammalian cell fusion was therefore often referred to as "*somatic cell mating*".

The original experiments on spontaneous fusion of mammalian cells suffered from two serious technical difficulties that needed to be improved before this technique could be used on larger scale. First, spontaneous cell fusion is exceedingly rare so that very large cell cultures had to be set up in order to recover a few fused cell lines. Second, it was cumbersome to identify the few fused cells that were mostly rather difficult to distinguish from the overwhelming majority of the non-fused parental cells in the culture. Similar to a needle in a haystack, very large numbers of subclones had to be examined microscopically to find the fused cells by morphologic parameters such as multiple or large nuclei, etc. The first problem, i.e., making cell fusion a more frequent event, was solved rather soon. In 1962 two virologists, the Japanese Y. Okada and the American B. Roizman, independently realized that the well-known ability of certain viruses to induce syncytia in mammalian cell cultures was nothing other than cell fusion, and could perhaps be exploited for that purpose. After budding from its mammalian host cell, a virus particle usually is covered by a coat that contains elements of the mammalian cell membrane. In order to enter into a new mammalian host cell, many viruses deposit specialized proteins in their surface coat that can attach to proteins of the host cell membrane and induce fusion of the virus coat with the host cell membrane. These specialized proteins are termed fusion proteins. Sendai virus has a particularly efficient fusion protein causing extensive cell fusion leading to visible giant multinucleated cells, so-called syncytia, in mammalian cell cultures infected with Sendai virus. Okada tried Sendai virus to induce fusion between mammalian cells, and it worked, increasing the rate of spontaneous fusion by several orders of magnitude. Up to 10% of the cells cultured in the presence of Sendai virus were the products of cell fusion. The use of Sendai virus was soon adopted by others interested in the induction of cell fusion, for example by Henry Harris at the University of Oxford, who published in 1965 a paper together with his co-worker J.F. Watkins in which they systematically examined the conditions of virus-induced cell fusion, demonstrating what they called "*a form of artificial sexuality imposed on mammalian tissue cells*". As previous investigators, Harris envisaged the application of cell fusion mainly as a more rapid way of mammalian mating. An important improvement in this paper was the use of Sendai virus that had been inactivated by UV light, thus avoiding the complication of live virus replicating in their cultures. They also were the first to achieve fusions between tumor cells of two different mammalian species, human and mouse. Henry Harris thereafter used cell fusion between normal cells and tumors to study the genetic control of malignant cell transformation. The question he hoped to answer was whether the normal or the malignant phenotype was dominant or recessive in the hybrid, and which were the genes that confer malignancy or benignancy to the hybrid cell. The cell fusion approach, however, could not solve this problem and became obsolete in cancer research when it was eventually bypassed by the discovery of oncogenes by the methods of molecular biology.

Cell fusion, spontaneous or induced by Sendai virus, is a random event. As a result, fused cells recovered from a coculture of two parental tumor lines could be either derived by fusion of two identical cells of either parent or by fusion of two different parental cells, the real heterokaryons. Only the latter are of interest, and methods were required that permitted reliable selection of heterokaryons. A pioneer in the development of such cellular selection methods for fused cells was John Littlefield, working at Harvard University, Boston, USA. Littlefield made use of previous findings by Waclaw and Elisabeth Szybalski and co-workers at the University of Wisconsin who made the basic discoveries leading to the ingenious selection systems that are still in use today and were absolutely instrumental in the development of the hybridoma technique by Köhler and Milstein. The Szybalskis and colleagues found that when cultured tumor cell lines were exposed to the toxic guanine analog 8-azaguanine, most cells died but a few could be recovered that continued to grow in the presence of the toxic agent, they were 8-azaguanine resistant. Detailed analyses of such resistant cells showed that they lacked an enzyme, hypoxanthine-guanine phosphoribosyltransferase (HGPRT). In normal cells, HGPRT is involved in DNA synthesis by transferring a ribosephosphate group to the purine base guanine, generating guanosinemonophosphate which then is incorporated into the DNA. HGPRT cannot distinguish between guanine and 8-azaguanine, leading to the incorporation of the latter into the DNA followed by cell death. HGPRT-deficient cells are therefore 8-azaguanine resistant, but can neverthless survive because purine nucleotides are synthesized also by a *de novo* pathway that does not utilize preformed nucleotides. Indeed, DNA synthesis being essential to all cells, mammals have evolved several redundant pathways for the synthesis of nucleotides. The pathway involving HGPRT is only a minor one, a so-called salvage pathway required in case of problems with the the *de novo* pathway.

In order to select fused cells, Littlefield realized that one needed resistance to a second toxic reagent. As first described by S. Kit and co-workers, he used cells that were resistant to 5-bromodeoxyuridine, an analog of the pyrimidine base thymidine. In normal cells, thymidine is phosphorylated by the enzyme thymidine kinase (TK), and then incorporated into DNA. TK does not distinguish between thymidine and 5-bromodeoxyuridine, resulting in incorporation of the toxin into the DNA and death of normal cells. TK-deficient cells are resistant to 5-bromodeoxyuridine but nevertheless viable because TK-dependent pyrimidine synthesis is also merely a salvage pathway, similar to HGPRT-dependent purine synthesis.

In 1964, Littlefield published the classical paper describing the fusion of two mouse tumor cell lines, one resistant to 8-azaguanine and the other resistant to 5-bromodeoxyuridine, with subsequent selection of heterokaryons. Of course, the toxins 8-azaguanine and 5-bromodeoxyuridine could not be used for selection, since the heterokaryons inherited HGPRT from the TK-deficient parental cell, and TK from the HGPRT-deficient parental cell. The fused cells would therefore be sensitive to both toxins. Littlefield thus took advantage of another selection medium, the famous HAT (hypoxanthine, aminopterine, thymidine) medium. The toxin in this medium is aminopterine which poisons the *de novo* pathway of DNA synthesis so that in HAT medium cells essentially depend on their salvage pathways. Aminopterine inhibits the enzyme dihydrofolate reductase which reduces dihydrofolate to tetrahydrofolate. Tetrahydrofolate is then cleaved again to dihydrofolate, releasing several C atoms and methyl groups that are required for nucleotide synthesis. With the *de novo* DNA synthesis pathway interrupted, non-fused parental cells in the culture die because they lack either one or the other of the salvage pathways. They are HAT sensitive. Fused cells, in contrast, are HAT resistant. They survive in HAT medium by making use of the added hypoxanthine that is converted to purinphosphate by HGPRT, and by making use of the added thymidine that is converted to pyrimidine phospate by TK. This selection medium, essential as it is until today in cell fusion, was originally described and christened HAT in 1961 by Elizabeth and Waclaw Szybalski, who rarely are given proper credit for it. In 1969, Waclaw Szybalski complained in a letter to Boris Ephrussi that he, Ephrussi, had quoted Littlefield rather than the Szybalskis for inventing HAT selection in a review article in *Scientific American*. Moreover, in 1980 Elisabeth Szybalska wrote a letter to Georges Köhler complaining that he and Milstein had neglected to quote the Szybalskis in their famous first paper on the hybridoma technique of 1975. Littlefield, in contrast, knew to whom he owed the HAT medium, and quoted the original paper by W. Szybalski, E.H. Scybalska, and G. Ragni correctly. Using their selection technique he concluded that "*in the present study, the detection of far fewer hybrid cells than were found previously has been possible through the use of two clonal lines of mouse fibroblasts, each containing a drug-resistant marker*."

In immunology, cell fusion did not become popular until the early 1970s. Several groups then reported on fusions of antibody-producing myeloma cells with non-immunoglobulin-secreting cells such as fibroblasts or B cell lymphoma lines. The general result of these studies was that maintenance of the ability to secrete immunoglobulin was dependent on the fusion partner, it was maintained in the hybrids with the B cell lymphoma, but was lost in hybrids with fibroblasts. This type of result was not unexpected and did not raise much interest in the cell fusion approach for the study of immunological problems. This situation changed drastically when Milstein used cell fusion to assess a pressing immunological problem that had remained unanswered since its discovery in 1965, the problem of allelic exclusion. The Italian Benvenuto Pernis and the German Eberhard Weiler had found independently of each other that single immunoglobulin-producing B cells from a heterozygous animal produced only one of the two allelic forms of its antibodies. About half of the B cells would produce one allelic form, the other half would produce the second allelic form. Similar to many other genes, the genes encoding immunoglobulin polypeptide chains are polymorphic in mice, rabbits, and man. A handful of allelic forms exist with a few differences in nuceotide and amino acid sequence, so-called allotypes, which have been widely used to follow the inheritance of immunoglobulin genes. The monoallelic immunoglobulin production of individual B cells was utterly surprising as no similar case of allelic exclusion was known in nature, cells nearly always express both allelic forms of a gene. However, the findings of Pernis and Weiler nicely corresponded to the clonal selection hypothesis postulating that each B cell expressed only one type of antibody, even though it was diploid like all other somatic cells and thus carried two sets of antibody genes. Milstein reasoned that the mechanism of allelic

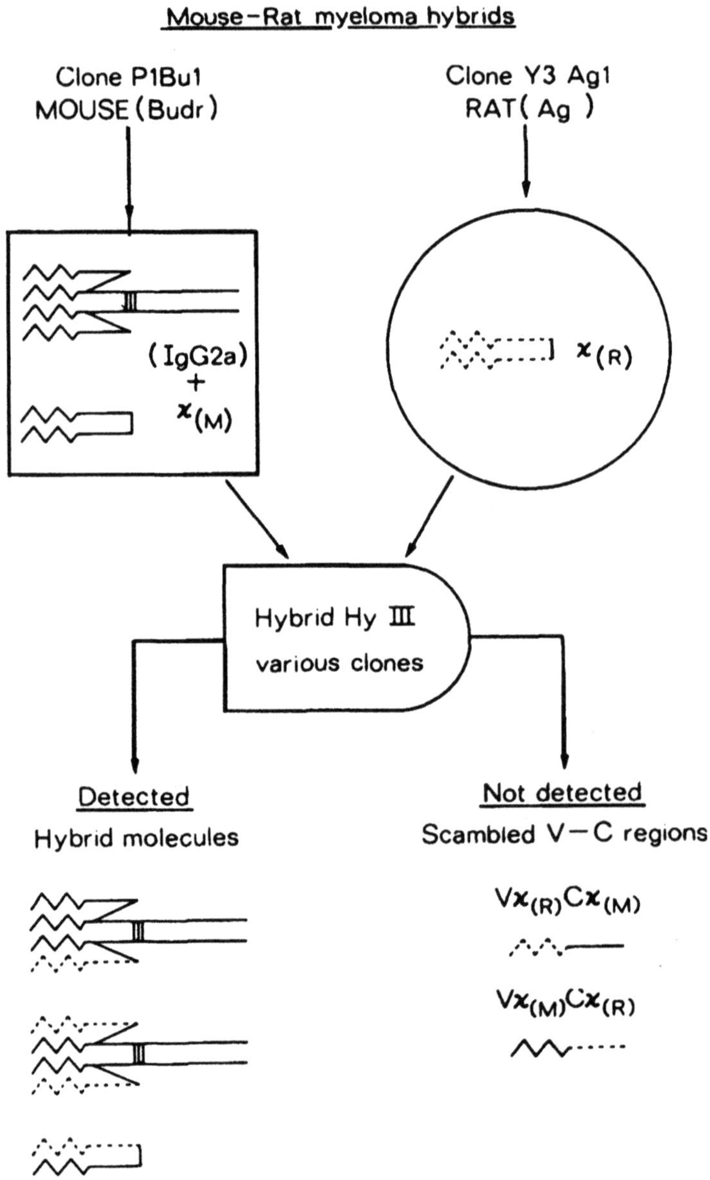


Figure 11
Schematic representation of the mouse-rat myeloma fusion of Cotton and Milstein. A bromodeoxyuridine-resistant clone of the mouse P1 myeloma was fused with an azaguanine-resistant clone of the rat Y3 myeloma. P1 cells produce complete IgG molecules and κ light chain dimers (solid lines), whereas Y3 cells produce only κ light chain dimers (dotted lines). Multiple clones of hybridization number III were analyzed for the composition of their immunoglobulin molecules. Hybrid molecules were found that contained mouse heavy and/or light chains together with light chains of rat origin in various combinations, but no mouse–rat V-C combinations were detected (from Milstein, 1984).

exclusion could be examined by the fusion of two cells that both secrete immunoglobulins. If a cell could not secrete more than one antibody, and if this was a fundamental law of nature, the hybrid cells should terminate production of one or the other parental myeloma protein, and the mechanism of termination should be amenable to investigation in such hybrids. Milstein and his collaborator Richard Cotton fused two myeloma cell lines, one of mouse origin and TK-deficient and the other of rat origin and HGPRT-deficient, and used HAT medium for selection of hybrids. The rat myeloma had been obtained from Herve Bazin, of Louvain, Belgium, who had developed the techniques to induce these tumors in the rat. The Sendai virus that Milstein and Cotton used to induce fusion was provided by their laboratory neighbor Abraham Karpas, who also taught them how to do tissue culture. Contrary to the expectations, the fused cells secreted both mouse and rat myeloma proteins, suggesting that a cell can readily produce two immunoglobulins if it contains the genes for two heavy and two light chains. Even worse, the heavy and light chains combined in all possible combinations resulting in novel types of immunoglobulins made of mixed mouse heavy chains/rat light chains and *vice versa.* Because the mixed molecules included asymmetrical versions as well, the hybrid cells produced an array of different immunoglobulin molecules, whereas normal B cells always make only a single symmetrical antibody consisting of two identical heavy and two identical light chains. Allelic exclusion in normal B cells was thus likely to be a process that resulted from the somatic generation of antibody genes, generating only one functional gene per cell for heavy and one for light chains. Another result of these experiments was also of interest. While the heavy and light chains of mouse and rat immunoglobulins reassociated randomly, there was no heterologous rat–mouse combination of V and C regions in either light or heavy polypeptide chains. This extended Milstein's previous result from mRNA sequencing in suggesting that the V and C genes had become integrated into a single VC gene already before the fusion, most likely at an early stage in the development of antibody-producing cells. Cotton and Milstein suggested that this may happen by some sort of translocation event, thus wisely anticipating the rearrangement process of antibody genes that was discovered only later by Tonegawa.

With these results Milstein visited the Basel Institute of Immunology in 1973 to give a seminar. At this occasion he met Georges Köhler who was just finishing his PhD thesis. The two agreed that Köhler should join Milstein's laboratory as a postdoctoral fellow, an agreement that eventually led to the invention of the hybridoma technique for the generation of monoclonal antibodies. As will be discussed in the next chapter, the crucial idea leading to this pathbreaking invention was to attempt fusion between a myeloma cell and normal B cells. It is noteworthy, however, that Köhler and Milstein were not alone in developing this idea. There was another group of investigators, Jerrold Schwaber and Edward Cohen at the University of Chicago, who at the same time did a very similar experiment to that of Köhler and Milstein, but lacking their focus and therefore remaining inconsequential. Schwaber and Cohen also wanted to understand allelic exclusion of antibody genes and fused the mouse myeloma TEPC-15, secreting antibody to pneumococcal polysaccharide, with normal human peripheral blood lymphocytes containing about 40% immunoglobulin-producing B cells. They used Sendai virus to promote fusion but did not use selection so that they had to search for fused cells by morphological means and only recovered a handful of hybrids. Importantly, these hybrids secreted, in addition to the mouse myeloma protein and mixed molecules, immunoglobulins produced by the human B cells. This finding could have been a direct line to the development of human monoclonal antibodies but the paper reporting the results did not contain any mention of such a perspective. The results were published in 1974, one year before Köhler's and Milstein's seminal fusion paper appeared.

## Chapter 6

# Köhler in Cambridge

Milstein and Köhler first met during a visit of Milstein at the Basel Institute of Immunology in 1973. The two seemed to have taken an immediate liking to one another, their personalities were somehow complementary in many ways. In Cambridge their personal sympathy developed into a friendship which included their families who frequently met privately in each other's homes. Köhler's relaxed attitude, quite critically commented upon by Melchers, was no problem to Milstein, it agreed with his principle: "*Laziness is the mother of good science. Creation comes from moments when you don't have anything to do*". Nevertheless, the overall atmosphere in the MRC was not without pressure to be successful at work, and coming to the laboratory on weekends or taking minimal vacations were matters of course. Unimpressed by this general attitude, during each of his two years in Cambridge Köhler took a four-week family vacation travelling through England with an old VW camper bus. This raised some eyebrows, if not with Milstein then with other scientists at MRC who felt that, rather than taking holidays, a young man should make the maximum use of a unique working opportunity. The Köhler family rented a nice house in the center of Cambridge and had active social interactions with the other members of the laboratory. The house contained a piano and Köhler tried to teach himself how to play it. Instead of taking lessons, he bought the notes for one piece of music, Chopin's Funeral March, and practiced playing it. After some time he would play it to entertain his guests, but it was the only piece anybody ever heard Köhler play.

Köhler arrived in Cambridge and began working in Milstein's laboratory in the MRC Institute of Molecular Biology in April 1974. In keeping with his reflective attitude to science, Milstein preferred to work with a small number of co-workers in his own laboratory, but collaborated closely with other group leaders at MRC, for example George Brownlee or Abraham Karpas. The latter had his laboratory next to Milstein's in the basement of the building and his people advised Milstein's group in the techniques of tissue culture. Köhler succeeded Richard Cotton who had just left to return to Australia, as a postdoctoral fellow in Milstein's laboratory. In addition, there was David Secher who was finishing his PhD thesis, and Shirley Howe, a senior technician and the heart of the laboratory who mastered most of the technical repertoire and was responsible for teaching it to newly arriving collaborators. Milstein and his co-workers, in looking for somatic mutations in B cells affecting the specificity of antibodies, had invested a lot of preliminary work into the subline of MOPC21, P3, that could be grown and cloned in culture and initially seemed a suitable model cell line for the project. First of all, they had determined the complete amino acid and nucleotide sequences of heavy and light chains. To analyze the immunoglobulins of P3 subclones, Milstein and Brownlee had established methods for radiolabelling the cells with the phosphorus isotope $^{32}P$ that was incorporated into the mRNA which then could be subjected to nucleotide sequencing. Moreover, they had worked out a method to label the secreted immunoglobulins by feeding the cells with a radioactive amino acid, $^{14}C$-lysine. In order to characterize the radioactive immunoglobulins, David Secher, then a PhD student in the laboratory, had set up the method of isoelectric focussing in gel slabs, much like Köhler in his PhD thesis in Basel. Other than Köhler, Secher visualized radiolabelled immunoglobulins by autoradiography. Köhler was very impressed with Secher's technology: "I was proud to be able to run four isoelectric focussing gels at a time in Basel. When I came to Cambridge, I saw that David ran 10 at a time". Using a soft agar cloning approach, Secher and Cotton were able to screen about 200 subclones of P3 within one week, mostly by the use of an isoelectric focussing device that was constructed in the MRC workshop and allowed to test up to 40 samples in one overnight run. After screening 2,000 P3 subclones the group had identified three P3 mutants secreting immunoglobulins with altered isoelectric charge. After screening a further 5,000 P3 subclones, partially after Cotton had already left, a total of seven such mutants had been collected. These mutants, however, were not what Milstein and his colleagues were looking for. They showed large deletions in the constant region genes rather than point mutations affecting the hypervariable region sequence (described in Chapter 2). Immunoglobulin genes with this type of mutation do not encode proper functional immunoglobulins and thus represent rare random accidents in DNA replication upon cell division. In the living animal cells with mutations of this type are rapidly eliminated by cell death, unless the mutation has activated a growth-promoting oncogene and turns the cell into a tumor cell. Cotton, in realizing that collecting P3 subclones may finally not yield the type of mutant that they wanted to study, reasoned that one would have to look for mutations affecting the specific antigen binding function of the secreted immunoglobulin rather than their overall isoelectric charge. Thus, a B cell line secreting antibody of known specificity was needed. To this end, Cotton had already started to look at mouse myelomas with known specificities, one of them MOPC 104E secreting antibodies to dextran. MOPC 104E grew in tissue culture, but extensive efforts to clone the cells, i.e., to grow up subcultures from single cells, failed. The most common reason for the inability to clone cells that can be grown as a mass culture is that the cells themselves secrete one or more factors into the culture medium required for their own growth. A single cell fails to grow because it does not secrete sufficient amounts of such factors. If this was the case for MOPC 104E is not known.

The other obvious way to obtain growing and clonable cells secreting antibodies of known specificity was to search for an antigen that was bound to the immunoglobulin secreted by P3. This was the project that Milstein initially suggested to Köhler. However, and in this case with good reason, Köhler refused, and Milstein wisely did not insist. The experiments would have to involve the testing of large numbers of different substances for reactivity with the P3 immunoglobulin, with only a small chance of success. The number of antigens to be considered was of equal magnitude as the diversity of antibodies, and the most likely outcome was that no fitting antigen was ever identified. Moreover, these experiments were likely to be tedious and boring. Instead, Köhler and Milstein agreed that he should continue the project already begun by Cotton, namely to try and establish clones and subclones from existing mouse myelomas with known specificity. After MOPC 104E could not be cloned, Köhler started to work with MOPC 315, secreting immunoglobulin binding to dinitrophenyl. The choice of this myeloma among several possible ones demonstrates some foresight as cells secreting antibodies to dinitrophenyl can be detected in a Jerne plaque test (see Chapter 2), after coupling dinitrophenyl to the sheep red blood cells. The Jerne plaque test is relatively easy to perform and allows the screening of very large cell numbers for secretion of specific antibody. A search for mutants that had lost or changed specificity would have been rather straightforward. However, MOPC 315 refused to grow in culture. Köhler tried for several months but could not get the cells to grow.

At this point Milstein was looking for a person who would continue Cotton's fusion experiments. Cotton had successfully generated fusions between mouse and rat myelomas, but his and Milstein's original aim had been mouse–mouse fusions. Cotton had obtained a bromodeoxyuridine-resistant variant of P1, one of the growing mouse myelomas in the laboratory. For fusion of two myelomas and selection in HAT medium, he needed azaguanine-resistant, HAT-sensitive variants of a second myeloma (see Chapter 5). He tried to obtain azaguanine-resistant variants of several myelomas including P3, of mouse origin, and Y3, of rat origin. An azaguanine-resistant variant that also died in HAT medium was first derived from the rat myeloma Y3, and so this was then taken for the fusion experiments with P1. These experiments provided two kinds of information: They showed that the hybrid cells produced the immunoglobulins of both parental cell lines, and that the V and C regions of the parental polypeptide chains were encoded by linked genes that stayed together in the hybrid and did not give rise to mixed rat–mouse polypeptide chains (see Chapter 5).

Nevertheless, Milstein wanted to see this result verified in a hybrid cell line derived from two mouse myelomas, and first asked Secher if he was interested in doing this experiment. Secher wanted to finish his thesis and so the task was given to Köhler, who was in search of a new project after his attempts to culture MOPC 315 had failed. Köhler thus continued Cotton's project by trying to make an azaguanine-resistant, HAT-sensitive variant of a mouse myeloma. As P3 had not worked in Cotton's hands, Köhler tried several cell lines including X63, a subclone of P3 that was originally isolated by Secher during his search for mutants with different isoelectric charge. During this study Secher had maintained many subclones with unchanged isoelectric charge because of other interesting phenotypes. X63 was maintained because it grew faster than the parental P3, and this might have been the reason why Köhler had no problem isolating an azaguanine-resistant variant, X63Ag8, from it. In fusing P1 with X63Ag8, Köhler obtained results that were in all aspects identical to those of Cotton's rat–mouse hybrids. Anything else would have been surprising, and Köhler's motivation to do these experiments in the first place remains open to speculation. Clearly, the experiments were not meant as pilot studies, preparations, or practicing exercises for the later hybridoma fusions. The co-dominant expression of both parental immunoglobulins in the myeloma hybrids was important and encouraging knowledge in this context, but the fact was known already from Cotton's results with the rat–mouse fusions. Nevertheless, the mouse–mouse myeloma fusions were Köhler's first successful experiments in Milstein's laboratory. Köhler considered these fusion experiments his primary project and continued them for several months in parallel with the first myeloma-spleen cell fusions (see below). They had two effects on Köhler: they gave him self confidence and they led him to think about cell fusions as a means for further experiments.

The question of how the concept of making hybridomas, i.e., of fusing myeloma cells with normal B cells of immunized mice, was born has been a matter of intensive and controversial debate. David Secher recalls that the idea developed between Köhler and Milstein "in the endless discussions that everybody always had with Cesar. It was somehow in the air, it was an obvious thing to try". Before monoclonal antibodies were realized to be one of the major scientific inventions of the century, intellectual priority was not an important issue, and the answer to the question can only be derived in retrospect by analyzing the controversy that arose after it became a matter of public interest. Several third-party scientists later made unproven claims to have suggested the experiment to Milstein, including colleagues as prominent as Sidney Brenner. It is documented that each of the two inventors, Köhler and Milstein, later thought of himself as having conceived the original idea and then persuaded the other to try it out in practice. While their discrepant opinions were initially kept mostly private and out of the public domain, the question became a pressing open problem several years later when the invention of monoclonal antibodies was up for science prizes and considered by numerous prize committees. Should Köhler as the young postdoctoral worker be included or should Milstein alone be honored? Milstein got under pressure in this public debate, and it became more and more unbearable to him after a number of important science prizes had been given to him alone. Should Milstein perhaps have insisted that the prize committees include Köhler? Köhler's and Milstein's rather close personal relationship became seriously compromised by Köhler's growing concern about not getting proper credit for his part in the invention. When in 1984 it came to the Lasker Award, though not connected with much money but highly prestigious and often a precedent to the Nobel Prize, Milstein suggested a meeting with Köhler to discuss how to deal with the problem. The result was a communiqué that left the question open but included Köhler as co-inventor of equal rights. It reads as follows:

*Statement by Georges J. F. Köhler and Cesar Milstein*

*Questions have been raised about the relative contributions which each of us have made to the design and execution of the experiments described in the paper published in 1975 in the journal, Nature, entitled "Continuous cultures of fused cells producing antibodies of predefined specificity".*
*We make the following statement:*
*We agree that both conception and execution of the work was the result of close collaboration between us with the skilled technical assistance of Shirley Howe. We are further convinced that the combined effect which resulted from such close collaboration was of synergistic nature. Synergistic effects taken to mean – as with monoclonal antibodies – effects that result from the combined action of two but cannot be produced by the two separately.*
*We both have a most pleasant memory of an exciting period in which a word, a comment or a passing remark made by one, had a resonant effect on the other. We do not want such happy memories which have sealed a close friendship to be disturbed by superficial interpretations of our individual recollections.*
*It was a collaborative work, it was a collaborative paper. We are happy to share this Lasker Award and we wish to make no further comments on this matter.*

*Signed, G. Köhler, C. Milstein*

Very appropriately, the 1984 Lasker Award included Michael Potter, the scientist who had produced the mouse myeloma tumors used in the fusions, in addition to Köhler and Milstein.

However, Köhler's opinion on his role in the invention was certainly not appropriately reflected by this statement and is better reconstructed by the evolution of the consecutive drafts of an article that appeared in the journal, *Science*, written by the science journalist Nicolas Wade, in February 1982. Wade originally wanted to write an article on monoclonal antibodies, their outstanding impact on science and medicine having become rather obvious over the years. He went to interview Köhler who was in the meantime back in Basel. After the interview he wrote a first draft of his article entitled "*How the hybridoma technique was invented*", including a verbatim quote of Köhler "*I was still thinking of trying to get a line which was specific for a given antigen. I had the idea in bed, before going to sleep, and then I couldn't sleep. I told Claudia, my wife, the next morning. Then I went to the lab, and I talked to Cesar, down in the basement, where the tissue culture was*". Obviously Köhler had very detailed memories about the circumstances in which it occurred to him that one should try to use normal B cells of immunized mice for fusion with myeloma cells.

Wade sent a preliminary draft to both Köhler and Milstein for suggestions. Celia Milstein recalls that her husband was deeply disturbed by the entire article, most of all by the way in which Köhler claimed priority in developing the idea. In a conference on Technology Transfer organized by the Wellcome Foundation in 1993 and primarily dealing with the patent scandal on monoclonal antibodies (see Chapter 8), Milstein recalls his reaction: "*When I read the account of Georges* (in Wade's draft) *about him having had a dream about this I was totally taken aback because until that moment I was completely convinced that the one who first thought about the experiment was me.... then I started to think how the two accounts can clash so badly. I have a very clear memory, having this conversation in the morning, next to the water bath, which was in the corridor incidentally.... I remember I suddenly had this idea and I said something like: Bloody hell, if we can't get a myeloma with antibody activity we should make it. I have that sentence in my mind*". And later in the discussion: "*This issue of who did the experiment was a bit traumatic for me because of this conviction I have had before and because I never said anything and because I did not want to contradict Georges at all*".

Milstein wrote the following response to Wade:

> *Dear Mr. Wade*
> *Thank you for the draft.... I am replying to you without having had much time to digest it properly, and to consider what my position should be.*
> *For the time being, I have decided not to get into any detailed discussion of the points you raise. Although I may not be happy to waste too much time on it, I do not mind discussing the "intellectual history of the discovery". But, most emphatically, I do not want to be drawn into any sort of public controversy of this matter. Not only do I have the highest regard for Georges, both as a scientist and as a friend, but I do not intend to cloud in any way the pleasurable memories and the intellectual excitement of our close interaction.*

*If, in spite of this letter, you decide to publish your draft, I may have to reconsider my position.*
*Yours sincerely*

*signed C. Milstein*

Milstein sent a copy of the letter to Köhler. In spite of being made aware of Milstein's strong disagreement with the article, Köhler made no corrections of the verbatim quote that Milstein had so obviously objected to. Later drafts of the article were not sent to Milstein who is cited incorrectly in all subsequent versions: "*Milstein declines to discuss the intellectual history of the discovery from his point of view....*" In contrast, Wade and Köhler closely communicated on several further versions of the article. Köhler's suggestions for corrections made his intellectual priority even more outspoken. A further advanced draft of the article was entitled: "*Hybridomas: World acknowledges only one inventor but there were two*". It contains an additional verbatim quote of Köhler: "*One of the things about Cesar is that he listens. If you come to him with a crazy idea, instead of dismissing it he will try to find out the good things about it...*" Thus, while giving Milstein credit as a considerate and interested supervisor, Köhler made again clear that he was the one who talked and Milstein was the one who listened in this case. Moreover, while the first draft had mentioned in a short final paragraph the US Dollar 100,000.– General Motors Cancer Research Foundation Prize shared by Milstein but not Köhler, later drafts and the published version contain one and a half columns on this issue. In these later versions Wade discusses at length two additional prizes (Horwitz Prize of Columbia University, Wolf Prize of the State of Israel) that had recently been given to Milstein, and criticizes the reasons for crediting Milstein, without including Köhler. Indeed, the issue of prizes finally became the main message of the published version, entitled: "*Hybridomas, the making of a revolution*" with the subheading "*Scientific prize committees sometimes skimp on their homework. The awards for the hybridoma technique may be a case in point*". Köhler, in contrast to Milstein, was fully in favour of Wade's article. For many years to come, Wade's article was the only document, besides his short curriculum vitae, that he used to send out if somebody approached him for information on the invention.

The debate about the relative contributions of Milstein and Köhler to the invention of the hybridoma technique is still actively going on today. In 2004 Sefic Alkan, former postdoc in the Basel Institute of Immunology and now at the Pharmaceutical Divisions in Minnesota, wrote a Landmark Article in *Nature Reviews Immunology* commemorating the 20th anniversary of the Nobel Prize to Jerne, Köhler, and Milstein. While the general message of the article was that all three scientists had contributed in their own way to the invention of monoclonal antibodies, Alkan had finished with the conclusion "*without Köhler we might have had to wait decades to put all this together*". Abraham Karpas, Milstein's laboratory neighbor and provider of the Sendai virus used in the initial fusion experiments, was sufficiently annoyed by Alkan's conclusion to submit a critique to the journal in which he recounted all the previous and subsequent work on cell fusion in Milsteins laboratory and concluded "*that in Sefic Alkan's article Georges Köhler's role ... is overemphasized*".

Both Köhler and Milstein agree that they had first discussed making hybridomas by fusing myeloma cells with B cells of immunized mice, one morning in the fall of 1974 in the corridor of the basement of the institute, next to the waterbath. Köhler then started to immunize mice with sheep red blood cells, an antigen chosen again because it was possible to detect cells secreting antibodies to sheep red blood cells by the Jerne plaque test. Because the spleen of immunized mice contains only very small numbers of B cells secreting antibody to the immunizing antigen, among a vast majority of B cells producing antibodies of unknown specificities, it was clear that large numbers of fused hybrid cells had to be tested, which was readily possible only by the plaque assay. Moreover, it was possible to detect specificity mutants by this assay. Alistair Cunningham, an Australian immunologist working in Canada, had developed a variant of the Jerne plaque test by using mixtures of sheep and horse red blood cells in the agar. B cells from a mouse immunized with sheep red blood cells would form two types of plaque, a clear type if both kinds of red blood cells were lysed by antibodies that crossreacted with horse red blood cells, and an opaque type if only the sheep red blood cells were lysed by non-crossreactive antibodies, leaving the horse red blood cells intact. In certain cases, plaques had a clear center and an opaque periphery, termed "sombrero plaques". Köhler was hoping to use Cunningham's technique to detect mutants of fused B cells changing from the crossreactive to the non-crossreactive antibody type, or *vice versa*.

Three myeloma cell lines were initially chosen as fusion partners: X63, a subclone of P3 originally isolated by Secher, 289, and P1. All three had been made azaguanine or ouabain resistant by Köhler in his attempts to fuse two mouse myeloma cells. Of advantage was that no second selection system was required for the normal B cells. Even though unfused normal B cells possess the salvage DNA synthesis pathways and are therefore in principle HAT resistant, they die in culture after a few days. Unfused drug-resistant myeloma cells also die in HAT medium, whereas the expected hybrids inherit the HAT resistance from the normal B cells and survive. Köhler set up his first fusion with spleen cells on 9 December 1974. Six parallel fusions were set up, using aliquots of $10^6$ and $10^7$ cells of each of the three myeloma lines together with $10^8$ spleen cells of mice that had been injected twice with sheep red blood cells. One day later the HAT medium was added and the medium was renewed on 11 and 13 December. On 17 December Köhler jotted in his notebook that all cells "*looked bad, both the presumptive hybrids and the P1 revertants*". Köhler split the cultures and added new medium again on 20 and 23 December, and the cultures recovered. However, his reference to "*P1 revertants*" indicated that in control cultures some of the P1 myeloma cells had spontaneously reverted to HAT resistance, a complication that meant that growth in HAT medium could not be used as firm evidence that fusion had indeed occurred. By looking at his cultures Köhler saw that he had cells growing in HAT medium, but he could not be certain that he had generated hybrid cells. What he then did with these cultures, or did not do, is again hard to understand: he just observed their growth in culture, splitting the cultures and adding new medium every few days, without analyzing them for antibody production. The Wade article quotes Köhler: "*I was reluctant to test them … because I thought I did not have specificity yet*", giving the impression that he thought it would take some time for hybrid cells secreting specific antibodies to arise and become detectable in the cultures. However, there was abolutely no sound reason to suspect a delayed appearance, or an enrichment with time, of such specific hybridoma cells. On the contrary, experienced cell biologists would have feared that hybrid cells with strong antibody secretion would perhaps grow a little slower than non-secreting hybrids and would thus get diluted out and disappear with time from the cultures. A rational approach would have been to test for antibody secretion as soon as possible and then to rescue the secreting hybrids, if any were found, by cloning. A more likely explanation is that Köhler's interest in the spleen cell fusions was limited and certainly not greater than in his myeloma fusions. He must have done the obvious calculations and come to the conclusion that it was pretty unlikely to find cells with specific antibody production among the hybrids generated by a single fusion experiment. Indeed, assuming that fusions between B cells and myeloma cells were random events, the expectation was that it would take many repeated fusion experiments to find a single hybrid secreting antibodies to sheep red blood cells. Also Milstein did not seem to have pressed Köhler to do the critical tests. Köhler later commented on the reason for his patience by saying: "I was content seeing the cells grow". Between 12 and 30 December Köhler's notebook contains multiple entries about experimental work on fusions between myeloma cells P1 and P3, and very little on his spleen cell fusions. On 30 December he finally labelled aliquots of the six spleen cell fusions by incorporation of the radioactive amino acid, $^{14}$C-lysine, planing to test them on 2 January by isoelectric focussing of the immunoglobulins secreted into the supernatant. However, the supernatants of the two fusions with X63, the only ones that later turned out to be successful, were contaminated with fungi and could not be analyzed. Köhler noted to repeat the labelling of the X63 fusions, while there are no entries on the results obtained on the fusions with 289 and P1. Most of January Köhler again spent with fusing myeloma cell lines. On 20 January he set up a cloning experiment of the X63-spleen cell fusions, while the 289- and P1-spleen cell fusions are no longer mentioned so that it is likely they had died out.

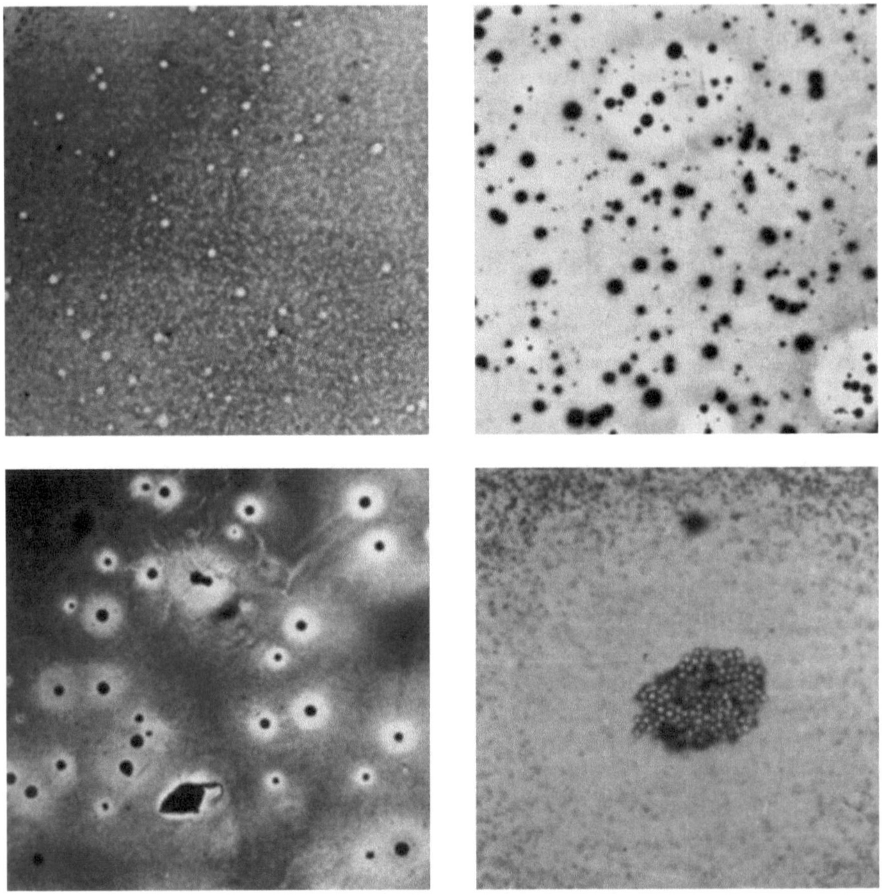

Figure 12
Jerne plaque assay of spleen cells of a mouse immunized with sheep red blood cells, and of hybridoma cells produced by fusion of these spleen cells with X63 myeloma cells. Each of the small white dots in the normal spleen cell assay corresponds to a single plasma cell secreting antibody to sheep red blood cells (top left, see also the magnification of a plaque in Fig. 7). Hybridoma cells plated in the same way grow up to small clones (top right, black dots). After addition of complement, circles of lysis of the sheep red blood cells are seen around many of the clones (bottom left, clear halos around the black dots). A larger magnification of a hybridoma clone in the center of a lysis halo is shown at the bottom, right (from Milstein, 1980).

Finally, on 24 January, Köhler made up his mind to do a Jerne plaque test on his X63 fusions. The event is described in detail in Nicolas Wade's *Science* article and is well known. Since plaques take 4–5 hours to develop, Köhler set up the assay in the afternoon and planned to return to the Institute after dinner to record the results. Expecting to see no plaques, he asked his wife, Claudia, to come along to the laboratory to keep him company while he looked at the negative plates. But then he saw the plaques, many of them. He became very agitated, kissed his wife, danced, etc. It was immediately obvious to Köhler that many of the hybrid cells secreted antibodies that bound to and lysed the sheep red blood cells in the agar. The frequency was orders of magnitude greater than calculated on the basis of random fusion between B cells and myeloma cells: 1 out of 400 cells in the fusion with $10^6$ X63 cells, and 1 out of 30 cells in the fusion with $10^7$ X63 cells formed a plaque, i.e. secreted specific antibody to sheep red blood cells. Later experiments revealed the reason for the unexpected high frequency. B cells secreting specific antibodies are in an activated state, whereas B cells that do not participate in the specific response to the antigen remain in a quiescent state and do not secrete their immunoglobulins. Activated B cells appear to preferentially fuse with the myeloma cells, perhaps because myeloma cells are also in a quasi-activated state and secrete their immunoglobulins. They appear to prefer their own lookalikes in the fusion event. Compared to the frequency of antigen-specific B cells in the spleen the enrichment of antigen-specific cells among the hybridomas is about 100-fold. This was totally unforeseen and is indeed the most important one of the many lucky events that came together to make Köhler's invention possible.

What followed was a very exciting period for Köhler and Milstein. The first fusion, termed SP(spleen)1, was cloned and clones of hybrid cells successfully isolated. The ability to obtain cloned hybrids was not at all a matter of course, but an important further prerequisite for the usefulness of the method. Two further fusion experiments (SP2, SP3) were performed and were as successful as SP1. In early 1975 a manuscript was prepared decribing the first three fusion experiments, all using sheep red blood cells as antigen. In addition to describing the fusion technique, it contained the data showing that the hybrids secreted the immunoglobulin polypeptide chains of both parental cells, and contained most or all chromosomes of both parental cells. Moreover, it demonstrated by the Jerne plaque test that some of the hybrid cells produced specific antibodies to sheep red blood cells, and that cloned hybrids could be obtained in nearly all cells which secreted specific antibodies. The paper was written by Milstein, entitled "*Continuous cultures of fused cells secreting antibodies to sheep red blood cells*" and signed by Köhler and Milstein. It was submitted in May 1975 to the journal *Nature*, published in Britain and among the most prestigious science journals. The manuscript was originally written and submitted as a "Nature Article", the leading one or two pieces in each issue, reserved for the most important discoveries and permitting detailed description of the findings at some length. *Nature* accepted Köhler's and Milstein's paper but did not find it of sufficient importance for publication as a Nature Article. The editors asked Milstein to submit a second version to be published as a "Letter to Nature", short pieces reserved for discoveries of lesser but still considerable import. It meant for Milstein to cut it down in length to 1500 words, which he reports to have been "a big problem". Moreover, for outstanding new discoveries published in *Nature* the editors usually invite another scientist of international standing to write a comment on the finding which is then published, along with the original paper, in the News and Views section of the issue. This was not done in this case. Nature editors certainly did not foresee the impact that this paper was going to have in the years to come.

When the paper was published in August 1975, the hybridoma technology was in the middle of a crisis of reproducibility. In the meantime, Köhler had successfully done fusions SP4 and SP5, both with sheep red blood cells as antigen. Did it also work for other antigens? In fusion SP6 and SP7 he used another antigen, trinitrophenyl (TNP) coupled to either lipopolysaccharide (LPS) or fowl (chicken) gammaglobulin (FGG). Similar to DNP, TNP is a small chemical compound that induces antibodies when coupled to a larger molecule, a so-called carrier, such as LPS or FGG. Of advantage, TNP could be used in the plaque test by coupling it to sheep red blood cells. SP6 worked well, showing that the method was not restricted to a single antigen. Only a few hybrids were derived in SP7 and thereafter, starting with SP8, the experiments ceased to work, hybrids were no longer obtained. Could it have been a streak of lucky circumstances, facilitated by unknown conditions that happened to be present in some cultures but were not repeatable in others? The experience that experiments sometimes do not give the results that had been obtained on previous occasions, under seemingly identical conditions, is well known to all biologists actively working at the bench. The number of unknown factors that play a role in experiments involving cell cultures is endless. Although careful investigators do their best to pretest new batches of fetal calf serum, media ingredients, antibiotics, sources of water, etc., many of these conditions cannot be controlled in every experiment. Milstein reports that they identified the source of the HAT selection medium as one reason in a later series of failed fusion experiments in Cambridge. The HAT ingredient is produced and sold by a number of companies in 10-fold concentration and is diluted 1:10 into the culture medium for selection of hybrids. After several unsuccessful fusions they found by accident that the particular HAT batch they were using was toxic, it killed not only HAT-sensitive cells but simply all cells. Experienced investigators may get annoyed but do not get discouraged in such a case.

Figure 13
Schematic representation of the hybridoma technology. Spleen cells of an immunized mouse are mixed with HAT-sensitive myeloma cells and fused by the addition of polyethylene glycol. Small numbers of hybridized cells are distributed into large numbers of microcultures and after a few days the supernatants of the microcultures are tested for antibody production. Microcultures that produce the desired antibodies are cloned on agar plates and clones are again grown in microcultures and tested for antibody production. Positive clones are recloned and frozen in liquid nitrogen. For mass production of the monoclonal antibody, hybridoma clones may be grown in bottles for collection of the supernatant, or they may be injected into the peritoneal cavity of mice where they induce ascites that contains the antibody. For reasons of animal protection, the latter procedure is banned in most developed countries (from Milstein, 1980).

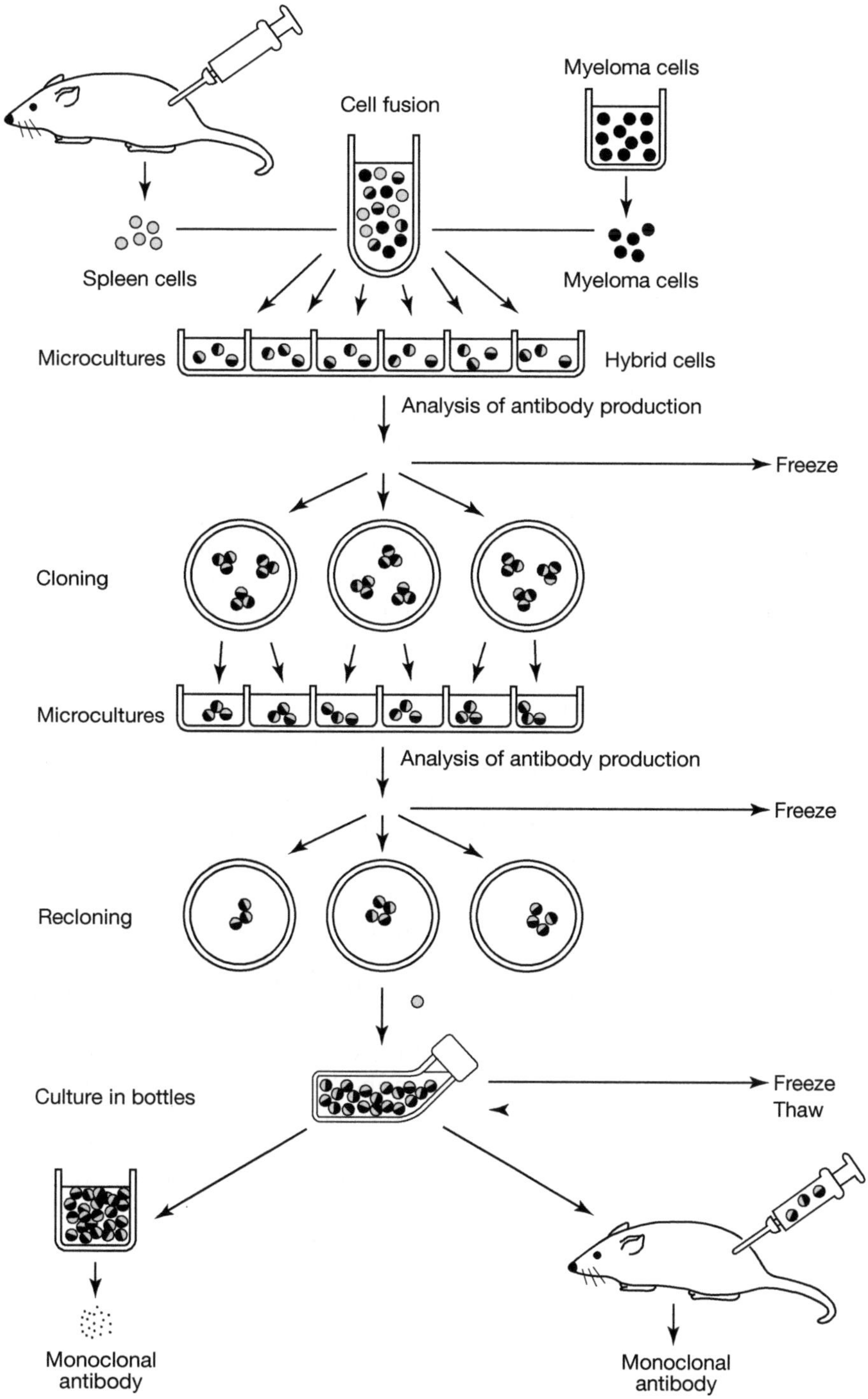
Myeloma cells
Cell fusion
Spleen cells
Myeloma cells
Microcultures
Hybrid cells
Analysis of antibody production
Freeze
Cloning
Microcultures
Analysis of antibody production
Freeze
Recloning
Culture in bottles
Freeze
Thaw
Monoclonal antibody
Monoclonal antibody

They either patiently test out conditions or simply throw away all reagents and prepare or purchase new batches of ingredients. Köhler's series of fusions started to work again at SP25. The reasons for failure were not identified in this case. Köhler later said that he became really convinced of the reproducibility of the method only after it worked in Klaus Rajewsky's laboratory, the first outside group successfully trying to repeat it.

For the rest of his two years in Cambridge Köhler was interested in the question how the hybrid cell forms antibodies having the choice of two heavy and two light chains. He showed that such a cell can make about a dozen different immunoglobulin molecules by combining each heavy chain with each light chain, including unsymmetrical molecules. Only some of these molecules are expected to have combining sites specific for the antigen, namely those in which the heavy and light chains of the normal B cell have been homologously paired. Pairings between the light chain of the normal B cell and the heavy chain of the myeloma, or *vice versa*, do not result in antigen specificity. Since unsymmetrical molecules are formed as well, some of them have only one instead of two antigen-specific combining sites, thus representing poor antibodies. Köhler came to the conclusion that good antibodies with two specific combining sites on the same molecule represented only a small proportion of the total immunoglobulins secreted by his hybrids. Thus, the hybridoma technique could possibly be improved by using a myeloma in the fusions that had lost the ability to secrete its own immunoglobulins. Loss of immunoglobulin secretion was not an uncommon event in hybridomas and Köhler detected several such cell lines in his fusions. One of those, termed NS (non-secretor) 1 and itself a hybridoma, was made azaguanine resistant and used by Köhler in fusion SP7. Upon testing clones derived from SP7, Köhler noted that NS1, even though it did not secrete immunoglobulin, still produced light chains that remained within the cell but became secretable when they could pair with a heavy chain in a hybrid cell. Nevertheless, hybrids from SP7 produced a less complex mixture of immunoglobulins with a higher proportion of good antibodies among them. The observation encouraged Rajewsky to search for an X63 variant that had lost the ability to produce both heavy and light chains, and Rajewsky's variant became subsequently distributed and used by most laboratories that produced monoclonal antibodies. By testing a number of hybridoma clones for what types of heavy and light chains they secreted, Köhler noted that hybrids lost the ability to secrete one or more of the four possible chains in a seemingly ordered fashion. He subsequently became interested in the phenomenon of chain loss, a subject that he intensively analyzed after his return to Basel (see next chapter).

---

Figure 14 (following pages)
The figure shows the pages of Köhler's laboratory notebook starting with the spleen cell fusion on 9 December 1974, and ending with the first analysis of the fused cells by plaque test on 24 January 1975. All scientists keep a notebook in which they enter all experimental steps in

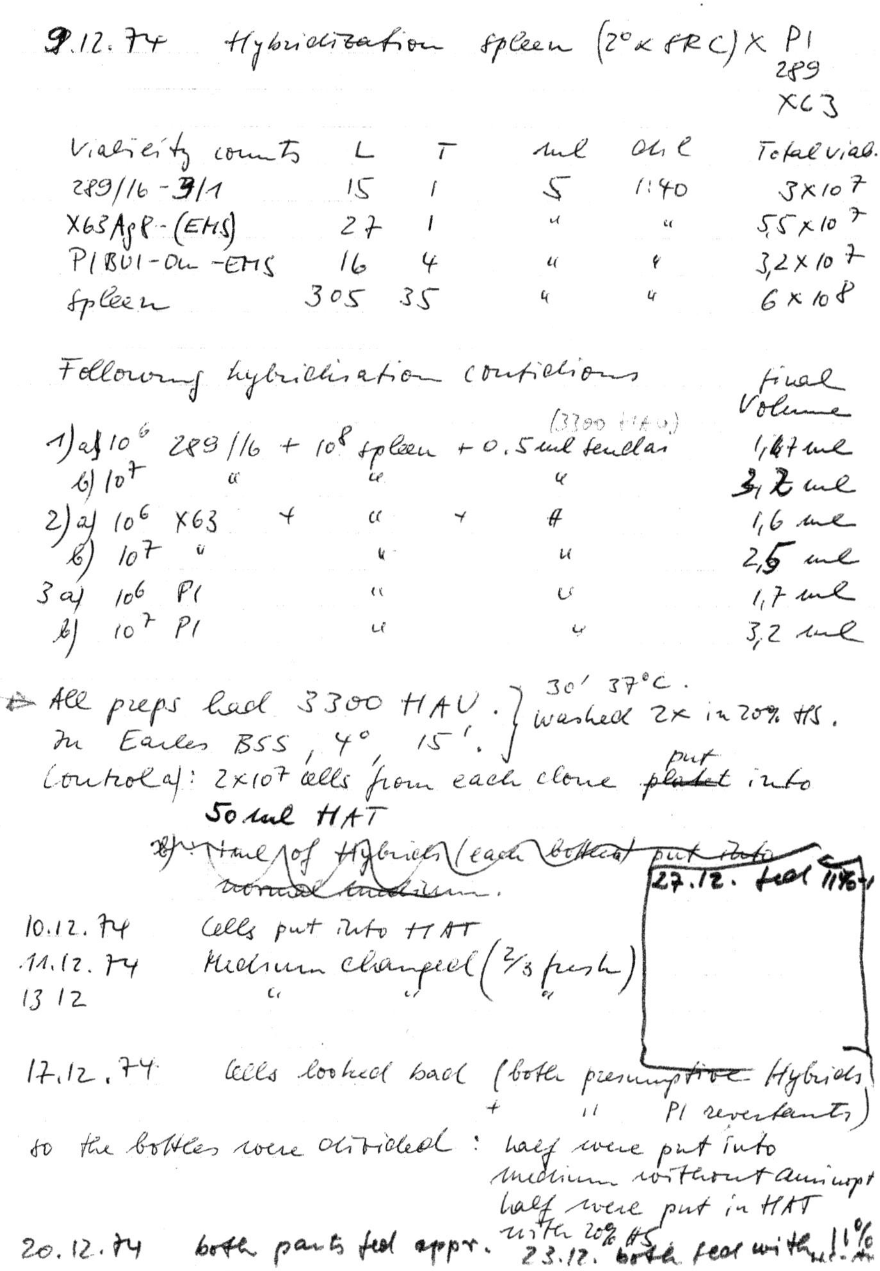
9.12.74 Hybridization spleen (2° x SRC) x P1
289
X63

| Viability counts | L | T | ml | dil | Total viab. |
|---|---|---|---|---|---|
| 289/16-3/1 | 15 | 1 | 5 | 1:40 | 3 x 10⁷ |
| X63Ag8-(EMS) | 27 | 1 | " | " | 5,5 x 10⁷ |
| P1BU1-Ou-EMS | 16 | 4 | " | " | 3,2 x 10⁷ |
| spleen | 305 | 35 | " | " | 6 x 10⁸ |

Following hybridisation conditions — final Volume

1) a) 10⁶ 289/16 + 10⁸ spleen + 0.5 ml Sendai (3300 HAU) — 1,67 ml
b) 10⁷ " " " — 3,2 ml
2) a) 10⁶ X63 + " + " — 1,6 ml
b) 10⁷ " " " — 2,5 ml
3 a) 10⁶ P1 " " — 1,7 ml
b) 10⁷ P1 " " — 3,2 ml

All preps had 3300 HAU. In Earles BSS, 4°, 15'. — 30' 37°C. washed 2x in 20% HS.
Controls: 2x10⁷ cells from each clone put into 50 ml HAT
~~Half of Hybrids (each bottle) put into normal medium.~~
27.12. fed 11%

10.12.74 Cells put into HAT
11.12.74 Medium changed (2/3 fresh)
13 12 " " "

17.12.74 Cells looked bad (both presumptive Hybrids + " P1 revertants) so the bottles were divided: half were put into medium without aminopt. half were put in HAT with 20% HS

20.12.74 both parts fed appr. 23.12. both fed with !!%

chronological order. Because scientists often do two or more series of experiments in parallel, entries even on the same day may relate to unrelated experiments. Most of Köhler's entries in the copied pages are concerned with fusions between different myeloma cells (courtesy of Deutsches Museum, Bonn).

12.12.74 | P1 × P3 Hybrids 1, 3, 6, 7, 8 + B have been cloned in non-areated bact. plates in medium supplemented with Thymidin + Hypoxanthin.
The following cell numbers were plated

| | | | |
|---|---|---|---|
| P1P3/1 | $10^5$ | $10^4$ | $10^3$ |
| P1P3/3 | $2{,}5 \times 10^5$ | $2{,}5 \times 10^4$ | $2{,}5 \times 10^3$ |
| P1P3/6 | $1{,}6 \times 10^5$ | $1{,}6 \times 10^4$ | $1{,}6 \times 10^3$ |
| P1P3/7 | $2{,}8 \times 10^5$ | $2{,}8 \times 10^4$ | $2{,}8 \times 10^3$ |
| P1P3/8 | $7 \times 10^4$ | $7 \times 10^3$ | $7 \times 10^2$ |
| P1P3/B | $2{,}7 \times 10^5$ | $2{,}7 \times 10^4$ | $2{,}7 \times 10^3$ |

All cells were grown previously in HAT medium

18.12.74 (4M Urea)
5-9 IEF-Run of Hybrids P1×P3
1: P1, 2: Hy 1, 3: X63Ag4, 4: Hy 3, 5: P1, 6: Hy 6, 7: X63
8: mixture P1 + X63, 9: P1, 10: Hy 7, 11: X63, 12: Hy 8, 13 P1
(20λ each: P1 + X63 from previous incorporation
radioautograph was developed 3 days)
18.12. same run but on 6-8 IEF
and (P1 Hy 8 P3) 2× strip overlay with
αP1 + αP3 sera diluted
1:10
(despite of Urea)

20.12.74 Cont. 12.12.74
All the Hybrids cloned very well ⇒ so too many clones even at the last dilution
therefore 12 clones of each put directly into 0.1 ml of *Lys-medium.
8 Hy 3 previously cloned (8) were also incorporated + P1 Col + X63 Ag 4 + Mix (supernat were frozen the 23.10. ⇒ all died

26.12.74 IEF (6–8)
with subcloned Hy 3 (P1×P3)

1.) P1 old
2. X old
3. 3–1
4. 3–2
5. 3–3
6. 3–4
7. mix old
8. 3–5
9. ~~3–6~~ mix new
10. 3–6
11. 3–7
12. 3–8
13. P1 new
15. Hy B
16. X new.

old: old incorp.
new: incorporation done same time

8.1.75 5× 1000 Hy B (P1×P3) plated in
5× 100 " normal medium

13.1.75 ~~now~~ no growth.

microzone section
barbitural pH 8.6
40'. 200 V.

283 (10⁶)
283 (10⁷)
X63 (10⁶)
X63 (10⁷)
P1 (10⁶)

X63

30.12.74 * Lys incorp. (0.2 ml)
2.1.74 supernatants stored in plate with Hy B

Hybridized with spleen cells
G1: 289 ($10^6$)
2: 289 ($10^7$)
3: X63 ($10^6$) } both contam. with fungi *
4: X63 ($10^7$) }
5: P1 ($10^6$)
6: —
7: X63 normal
8: " " } new incorporation
10 X63 ($10^6$)
11 X63 ($10^7$)
H1: Hy 3 (289 x P1)
H2: Hy 4 ( " ) contam. with fungi *
H3: —
H4: 289 normal
H5 P1 "
H6 289 + P1 mix

2d IEF (6–8) run (20 λ)
1. 289
2. Hy3 (289xP1)
3. P1 (40 λ)
4. Mix ← 5 289
*6 ~~5~~ Hy 4 (289xP1)
7 ~~6~~ P1
8 ~~7~~ 289
9 ~~8~~ 289 ($10^6$)
10 ~~9~~ 289 ($10^7$)
11 ~~10~~ X63
* 12 ~~11~~ X63 ($10^6$)
* 13 ~~12~~ X63 ($10^7$)
14 ~~13~~ P1
15 ~~14~~ P1 ($10^6$)

Fusion 15.1.75

| | L | | Total |
|---|---|---|---|
| X63 : | 1,5×10[5] /ml | 20ml | 3×10[6] |
| P1BU1 : | 1,3×10[5] " | | 2,6×10[6] |
| P1BU2 : | 3,7×10[5] " | | 7.4×10[6] |
| 289 : | 3,2×10[5] " | | 6,2×10[6] |
| Pr4 : | 1,9×10[5] " | | 3,8×10[6] |
| D1/20 : | 4×10[5] " | | 8×10[6] |

following hybrids were done (each 10[6] cells + 1ml of Sendai ≙ 4000 HAU).

1.) P3 × P1BU2 } controls: of each strain one drop in normal
2.) 289 × P1BU2 } medium + 1ml ≙ 10[6] cells in BSS mixed with 1ml HAT.

3.) Pr4 × P1BU1 (8×10[-4]M Ouab.) } same controls as above
4) D1/20 × P1BU1 " } but HAT is supplemented with 3×10[-4]M Ouabain.

after 15' at 4° and 30' at 37 the cells were washed twice in normal (11% HS).
and distributed in 8 wells (1+3) and 10 wells (2+4).

16.1.75 addition of selective medium
17.1.75 change of ½ the medium.
20.1.75 change
24.1.75 " the D1/20 × P1 was fed with 6×10[-4]M Ouab. instead of 3×10[-4] because D1/20 didn't seem to be too much affected.

28.1.75 "

21.1.75 a) IEF 5–8 w/o Urea (40λ) 289/116 Ig 411
40λ of the supernatant (320λ) after ultracentrifuge run. (cell had 1h × Lys!) intracellular heat.
289/116 × P1BU1 – 
" " 2

P1BU1 Orab
P1BU2

each sample ~~the gel~~ was overlayed with a paperstrip containing ~ 0.2–0.3 ml of spec. α P3 (left) and spec. α P1 (right), undiluted. Overnight, RT in moist box. The strips were removed after 22 hours. The gel put into 500 ml of PBS for 1h at RT.

b) Hy B (20λ) was run in a 6–8 pH, 4 M Urea gel, the gel was cut + put into a 2% mercaptoethanol solution (300 ml) PBS, RT) for 15 min. Then the acrylamide strip was removed from the glass + put onto a new 6–8, 4 M Urea plate.
(the problem is to see the original lines before ME-treatment).

c) pH 5-8, w/o urea gels with 30λ
P3, Hy B, P1 on 0.5 x 3cm 3MM Whatman paper.
~~xxxx~~ αP3-strips (left) + αP1 strips (right) (see under a)) over night.
Same for:
X63 x spleen (10⁶), X63, X63 x spleen (10⁷).

23.1.75 the supernatants from (a) 21.1.75) + X63-Ag8 were washed 3x in PBS ~2000 rpm little forwald in cold room swinging bucket 15'. The precipitate was taken into 50λ 8M Urea 4% Ampholytes pH 3-10. After a few hours solid Urea was added to saturate the solution. Left over-night, RT. ⇒ the solutions are clear.
20λ of X63, 2x3, 3, 4, P1, BU2 (left)

the rest (40λ) was given 2λ of Mercaptoethanol and incubated 15' at 60°. and loaded in the same order on right hand side of the gel

24.1.75 X63 Ag8 (10⁶) x 10⁸ spleen Test
X63 Ag8 (10⁷) "
for α SRC plaques.
Normal SRC were mixed with 2,4x10⁴ X63 (10⁶) and 3,8x10⁴ X63 (10⁷) cells / plate incubated at 37° in Hotpack for 4h and then developed for 3h ⟶

with: N° of plaques

| | X63 $10^6$ | X63 $10^7$ |
|---|---|---|
| 2nd lilo C' alone | (4) clear plaques. 100 (very weak + small) | 1388 small but clear plaques. |
| R α IgG2a (1:100) sheep α IgA(315) 1:20 + C | 64 (small, clear) much better than above | 1256 (small, clear) |
| + sheep C8 α Mouse P3 1:10 + S α IgA 315 (1:20) + C | 57 (small, clear) | 1376 (smaller than above, but clear) |

conclusion: a) 60 cells in $2,4 \times 10^4$ cells (X63 $10^6$) are forming indirect SRC plaques. ≅ 1/400

b) 1300 cells in $3,9 \times 10^4$ (X63 $10^7$) are direct plaques ≅ 1/30

# Chapter 7

# Back in Basel

Köhler returned to Basel to become a member of the Basel Institute of Immunology as of 1 April 1976. As planned by Niels Jerne, the hierarchy at the Basel Institute had initially only four levels, the Director, the scientific members, PhD students, and the technical staff. With some notable exceptions, members worked by themselves, only supported by a technician and in some cases by a graduate student. The position of postdoctoral fellow, i.e., a junior scientist working under the supervision of a senior scientist, did not exist in Jerne's system. All postdoctoral scientists, experienced or not, designed and carried out their own research programs independently. The funds to finance the program were given without prior formal evaluation. Köhler, at age 30 and with two years experience as a postdoctoral fellow and still a rather junior scientist, was very clear about what he did not want to do. At numerous later occasions he recalled his main fear at the time: "I did not want to become a monoclonal antibody maker". In other words, he feared that colleagues and perhaps superiors at Hoffmann-LaRoche would press him to generate monoclonal antibodies against antigens of their interest, a task which would have made him a service scientist for multiple projects of no immediate interest to him. He much preferred to continue trying to identify the mechanism of immunoglobulin gene diversification and, as inventor of hybridomas, thought that he had much better tools at hand than his competitors. He believed that hybridomas, as they grew in tissue culture going through many generations of cell division, would mutate the genes encoding the antibody polypeptide chains. By isolating such mutants he hoped to be able to determine the mutation frequencies and by analyzing the genes he hoped to identify the types of somatic mutations that diversified antibodies. His rationale was rather similar to that of David Secher in Milstein's laboratory when isolating mutants of the P3 myeloma cell. Köhler and his colleagues with whom he discussed his plans must have been aware of Secher's failure to isolate mutants affecting the hypervariable regions of the P3 myeloma, strongly suggesting that once a plasma cell has turned malignant it ceases to diversify its antibody genes. At what stage in differentiation B cells stabilized their antibody genes was unknown, however, and it was possible that non-malignant plasma cells still mutated. However, since in the generation of hybridomas normal plasma cells are fused with malignant plasma cells, it was not unlikely that the hybridoma behaved like a myeloma. More likely than not, mutations would have stopped in hybridomas as well. Disregarding this risk, Köhler collected mutants of mostly one hybridoma clone that he had brought to Basel from Cambridge, SP6 with specificity to TNP. Like Secher and others before him, most of the mutants he found were chain loss mutants, i.e., variants of SP6 that had stopped to produce one or more of the four immunoglobulin polypeptide chains that hybridomas produce. Somewhat opportunistically, he became interested in this most frequent type of variant and, upon recording the frequencies and order of chain loss, made an intriguing observation: chain loss was not random, it was governed by certain rules. The most frequent event was that hybridomas producing two heavy chains and two light chains lost one of the heavy chains whereas the loss of one of the two light chains was somewhat rarer. In a hybridoma that had already lost one heavy chain and one light chain, loss of the second heavy chain was again frequent whereas loss of the second light chain never occurred as long as a heavy chain was still produced. However, once a hybridoma had lost both of its heavy chains, light chain loss occurred with high frequency. He thus concluded that hybridoma cells could not survive if they produced only heavy chains but had no problem surviving with just producing light chains. In other words, free heavy chains, not paired with a light chain, would kill the cell, unlike free light chains and complete immunoglobulins consisting of both heavy chains and light chains. He further studied this phenomenon by so-called hybrid hybridomas that produce three or four different immunoglobulin chains, and finally proposed what became known as Köhler's "heavy chain toxicity hypothesis". On the basis of this hypothesis, Köhler developed a theory on the development of B cells in which heavy chain toxicity played a major role. It had just been found by others that in the course of development, B cells rearrange their heavy chain genes some time before their light chain genes, and thus express heavy chains without light chains for some time. Köhler proposed that during this time B cells are under pressure to rapidly produce light chains that pair with the heavy chain to neutralize its toxicity. Only B cells that manage to do so can survive. In Köhler's theory this was the way to guarantee that finally all B cells produce a pair of heavy and light chains. While heavy chain toxicity in B cells may indeed exist, subsequent detailed investigations of B cell development have not revealed an important role of this mechanism. Fritz Melchers, of all his colleagues, was the one who showed that the heavy chain in early B cell development is not alone. Until proper light chains are produced by rearrangement of the light chain genes, the heavy chains pair with so-called surrogate light chains, invariant light chain-like proteins that do not require gene rearrangement. Heavy chain and surrogate light chain together form the so-called pre-B cell receptor on the cell surface. Survival of the B cell during this period depends on signals coming from the pre-B cell receptor, i.e., survival is not endangered but rather depends on the heavy chain.

While doing the experiments on heavy chain toxicity, Köhler did not forget his interest in finding mutants of hybridomas that affected the specificity of the secreted antibodies. He had learned in his search for loss variants that if he screened random subclones of SP6 in the Jerne plaque test and looked for those that did not lyse sheep red blood cells coupled with TNP, the vast majority ot them did not secrete any antibody at all. In order to detect the elusive specificity mutants, apparently much rarer, he would have to screen thousands of random subclones to come up with a few mutants that secreted antibodies that did not bind to TNP, just like Secher had done earlier looking for P3 mutants with altered isoelectric charge. This was not what Köhler considered an intelligent approach. According to Milstein's principle that laziness is the mother of good science, Köhler set out to develop selection systems that would allow him to enrich mutants in culture by first getting rid of all non-mutated SP6 cells. The principle that he used was based on the fact known from the plaque test, namely that antibodies to TNP killed cells that were coated with TNP on their surface. He thus coated his SP6 hybridoma cells with TNP before he placed them in culture. There the hybridoma cells started to secrete antibodies, the antibodies would bind the TNP on the hybridoma cell surface and kill the cell. The procedure had to be repeated a few times but then only mutant cells were left in the cultures. The idea seemed ingenious, non-mutated hybridoma cells secreting antibodies binding to TNP would commit suicide in culture, while mutated hybridoma cells would survive. However, the suicide selection system was not quite as efficient as Köhler would have liked it to be. Cells surviving the suicide selection included all non-secreting hybridoma cells, as well as cells that secreted antibodies that bound to TNP but were unable to lyse the hybridoma cells. Suicide killing depended on a serum factor, complement, that is added to the cultures and facilitates cell lysis by binding to the constant region of the antibody heavy chain. As a consequence, heavy chain constant region mutants with defective binding of complement were also included in the hybridomas surviving suicide selection. Nevertheless, the system worked and Köhler isolated a number of mutants. However, when he wanted to analyze the mutated antibody genes, he faced a problem. Niels Jerne did not permit him to analyze DNA. There was another scientist in Basel, Susumo Tonegawa, who had claimed for himself that he was the only one to analyze mutated antibody genes at the Basel Institute.

When Köhler arrived in Basel in 1976, Tonegawa had just published his pathbreaking experiment showing that antibody V- and C-genes, while being located in separate clusters in the genomic DNA, rearranged in B cells such that a single V-gene moved close the the C-genes so that a continuous polypeptide chain, consisting of V and C region, can be synthesized. There were few who did not realize the importance of this experiment so that Tonegawa had a rather strong position in the Basel Insitute. He now wanted to study the still elusive mechanism of somatic mutations and took an approach very different from that of Köhler to attack the problem. Several laboratories worldwide were engaged in the sequencing of antibody genes and the structure of the antibody loci became more and more known. One of the surprising findings of these efforts was that V-genes were not continuous in the genome but came in several parts that had to be rearranged in B cells to form a continuous V-gene (see Chapter 1). For example, the light chain V-genes in B cells were composed of two genomic sections, V and J. Of the two classes of light chains, κ is the major and λ is the minor one in the mouse. Light chains of λ class fall into two V region subgroups , λ1 and λ2. Most importantly, Tonegawa found that the BALB/c mouse, the strain from which most of Potter's myelomas were derived, had only a single V-gene for λ2 light chains in the genome, coding for amino acids 1-96 of the λ2 V region. In contrast to κ, where it was not known how many still unknown V-genes were to be discovered in the genome, all λ2 light chains of the BALB/c mouse must be derived from that single Vλ2 gene in the genome. Thus, all that Tonegawa had to do was to analyze the DNA sequences of BALB/c myelomas that had light chains of the λ2 class, and compare the sequences with the known genomic Vλ and Jλ sequences. Any differences must have resulted from somatic diversification events. Martin Weigert and Melvin Cohn had previously sequenced some myeloma λ light chains that belonged to the Vλ1 V region subgroup, and had postulated somatic mutations to explain their results. However, this was before it was known that there was only a single Vλ1 gene as well. Tonegawa, having this knowledge, was able to interpret his results in a much more definitive fashion. Sequence differences within the genomic V-gene could only be somatic mutations, sequence differences in the joining region of V and J would result from variations in the recombination event.

Not only because he was a molecular biologist, Tonegawa chose to sequence DNA rather than protein. Obtaining a protein sequence of an entire light chain was still an endeavor of well over a year, if all went well. The technique of DNA sequencing was faster, but nevertheless was still in its infancy and also a difficult technique to master. In particular, if one wanted to be sure that a single nucleotide exchange was a point mutation and not an error in sequencing, one had to repeat a sequencing step several times. Foreseeing these difficulties and facing all these obstacles, Tonegawa demanded of Jerne to not allow any other laboratory to do DNA analyses in the Institute. Tonegawa was supported in his demand by Charles Steinberg, who had intensively counselled Tonegawa in planning his experiments. Like Steinberg, Jerne must have been convinced that Tonegawa's approach to understanding somatic diversification of antibody genes was the one with the fastest and greatest chance of success, and complied.

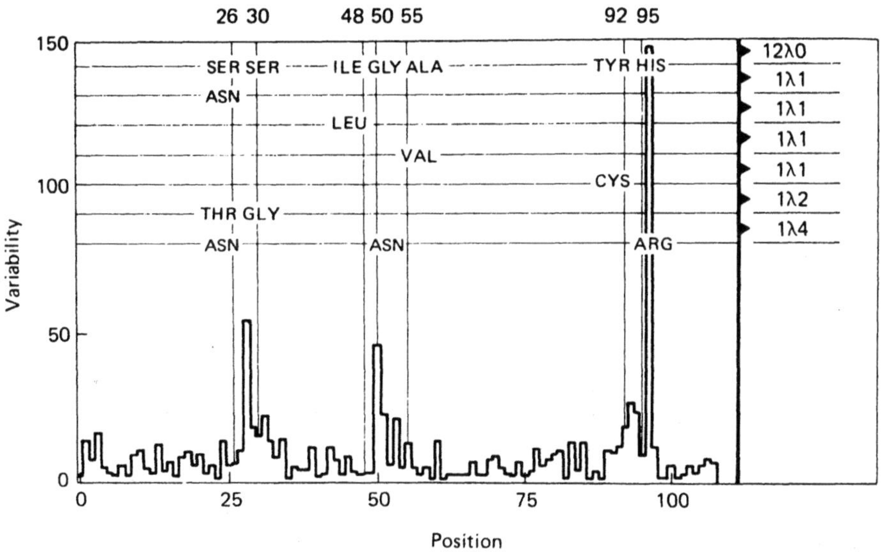


Figure 15
Mutations in myeloma light chains of the λ1 subgroup, superimposed onto the Wu-Kabat variability plot of all light chains. Among 18 light chains analyzed, 12 had identical sequences (then termed λ0. The numbering in this historical figure does not correspond to the present subdivision in λ1 and λ2 V region genes). The additional light chains showed 1, 2, or 3 unique amino acid differences at the indicated positions. Note that all differences occurred in the hypervariable regions. The authors proposed that the unique amino acid differences were due to somatic mutations of a germ-line V-gene corresponding to the λ0 sequence. They were correct but could not be certain because it was not known at the time that there was only one V-gene for each λ subgroup (from Weigert et al, 1970).

Indeed, Jerne's hunch was correct. Whereas Tonegawa made major contributions to the understanding of antibody gene diversification by analyzing λ light chain sequences, Köhler's hybridomas contributed little to the subject. However, before finding this out, Köhler had to first find a laboratory in which he could analyze the DNA of his mutant hybridomas. He found it in the European Molecular Biology Laboratory (EMBL) in Heidelberg, where he was accepted as a guest in the group of Hans Lehrach. He thus drove from Basel to Heidelberg several days a week to do his analyses there instead of his own laboratory. According to Melchers, Köhler accepted Jerne's decision and continued to be on speaking terms with both Jerne and Tonegawa.

Köhler had collected approximately 20 mutant lines by the suicide selection approach from two hybridomas, SP6 and PC 700, the latter secreting antibodies to phosphorylcholine (PC). The mutants secreted antibodies but did not lyse sheep red blood cells coated with TNP or PC, respectively, in the presence of complement. Disappointingly, mutations leading to loss of specificity by altering the sequences of the hypervariable regions of heavy or light chains were not found in his collection. Instead, he identified a number of deletion mutants in which entire DNA exons encoding individual domains of the constant region of the heavy chain were deleted. In addition, frameshift mutations were found resulting in premature termination of translation and severe shortening of the heavy chain. In other words, the mutations that Köhler isolated did not represent the somatic diversification mechanism he was hoping to study. Rather, the mutations he described were similar to the P3 mutations observed earlier by Secher and Milstein. They just represented the occasional mistakes made by all cells upon DNA replication and had no physiological meaning. Whatever the somatic diversification mechanism in B cells was, it did not exist in hybridomas. Marc Shulman, a Basel colleague who had become interested in Köhler's mutants and had participated in analyzing them in Heidelberg, was deeply disappointed and expressed this in letters which he wrote to Köhler during the preparation of a manuscript on these results. In contrast, Köhler looked at his mutants rather stoically and tried to make the best of them. For example, the heavy chain domain deletions could be used to map the epitopes recognized by a series of monoclonal antibodies to IgM produced by the Melchers group, as well as for the mapping of the glycosylation sites and the complement binding sites on the heavy chain. Later, already in Freiburg, Köhler used his deletion mutants for studies on the pre-B cell receptor and gave them to Richard Lynch who used them to map interactions of IgM with IgM receptors on lymphocytes.

Were hybridomas totally useless for analyzing antibody gene diversification? They were not, one only had to use them in the right way. Milstein, realizing that hybridomas had stabilized their antibody genes, exploited exactly this property to attack the diversification mechanism. He reasoned that one could stabilize B cells that were mutating their antibody genes upon responding to an antigen *in vivo*, by producing a series of hybridomas from splenic B cells of immunized mice at various times during an immune response. By this means, single B cells could be stabilized at any phase of an immune response, expanded and immortalized to yield lots of cloned hybridoma cells for any type of analysis one could think of. Analyzing the V-gene sequences in such hybridomas would reveal any type of mutation, if they had occurred. By that time the method of mRNA sequencing had been developed to allow the determination of nucleotide sequences reliably and rapidly, so that large numbers of hybridomas could be analyzed in a reasonable time frame. For an antigen he chose oxazolone, a small chemical not unlike TNP that induced strong antibody responses when coupled to a protein carrier. Together with his postdoctoral fellows Matti Kaartinen, Claudia Berek, and Gillian Griffith, Milstein made several fundamental discoveries in this system. They found that until 7 days after immunization with oxazolone the vast majority of B cells produced antibodies that were encoded by single non-mutated genomic V-genes, one for the heavy chain and one for the light chain. There was some junctional diversity at the VDJ or VJ junctions of the V-genes but there were no mutations within the V-genes. In sharp contrast, mutations had occurred at day 14 after immunization, for the most part single nucleotide exchanges in the hypervariable regions of the V-genes. After injection of a second dose of antigen the mutated B cells of the day 14 response appeared again in the hybridoma population but there were also new B cells expressing mutated forms of other genomic V-genes. Antibodies that were encoded by the mutated V-genes differed from that encoded by the non-mutated V-genes by a higher affinity to oxazolone, i.e., they had a better fit to the antigen. These data were the first to demonstrate somatic mutations of antibody V-genes occurring during an immune response *in vivo*. They also provided a rational explanation for the long known phenomenon of affinity maturation, i.e., the appearance of better antibodies with time during ongoing immune responses. The results for the first time allowed rough estimates of the mutation frequency of antibody genes which was found to be a thousand-fold greater than in bacteria. Thus, B cells seemed to have a mechanism of somatic hypermutation, creating many more mutations than expected with a spontaneous mutation rate. Thus a new avenue of research was born, aimed at understanding somatic hypermutation. Not much later, Rajewsky and his collaborators did similar experiments and obtained similar results on the immune respone to nitroiodophenacetyl (NIP), revealing the general nature of the observations. Moreover, they found that somatic hypermutations occurred at defined anatomical sites in the spleen, the so-called germinal centers, where B cells carrying antibodies as receptors are first exposed to the antigen, then meet with the T helper cells to take a bath in their cytokines, and finally proliferate before they differentiate into plasma cells and secrete antibodies.

The detection of somatic hypermutation of antibody genes during immune responses would not have been possible without hybridomas. Somewhat tragically Köhler had no part in these exciting studies, although this had been the ultimate motivation behind his invention. The reasons why he did not pick up these ideas are not clear. Although Köhler's relationship to Milstein had become temporarily disturbed he kept in close and friendly contact with Rajewsky so that he must have been exposed to the concepts behind their experiments. On the one hand, it is possible that experiments of such complex design as that of Milstein and Rajewky did not agree with Köhler's more straightforward way of thinking. On the other hand, elaborate experiments of this type required institutional support with considerable resources including manpower, conditions that were not available to Köhler in Basel. Nevertheless, Köhler was among the first to make use of hybridomas in another most fruitful area of research, i.e., as a source of antibody genes for the construction of transgenic mice. Transgenic mice became a major tool of investigation in immunology in the years to come, and Köhler had a leading part in their development (see Chapter 13).

Meanwhile, an ever increasing number of laboratories worldwide had begun to produce monoclonal antibodies for use as reagents in research and medicine. Immunologists realized quickly that monoclonal antibodies were ultimately superior as reagents to the conventional antisera that had been used so far. In addition, making monoclonal antibodies was not much more difficult than making conventional antisera. The first to employ the technique for practical antibody production were British scientists, including among others Alan Williams and Jonathan Howard, who worked together with Milstein to generate monoclonals to cell surface antigens for use in blood cell typing, blood group determination, and transplantation typing. Soon afterwards the Americans caught on. Timothy Springer at Harvard University and Len and Lee Herzenberg, scientist couple and pioneers in flow cytometry at Stanford University, made whole collections of antibodies against cell surface antigens for differentiation of mouse and human leukocytes. These antibodies promised to become indispensable tools in basic research as well as in diagnostics of lymphomas and leukemias. Hilary Koprowsky, director of the Wistar Institute in Philadelphia, was the first to produce a monoclonal antibody to a human melanoma tumor, raising hopes for its possible use in anticancer therapy. Not much later Koprowsky reported on monoclonals against rabies virus that were useful as typing reagents to distinguish virus subtypes. David Secher, a former student in Milstein's laboratory, had managed to produce a monoclonal antibody to the antiviral substance, interferon-α, and succeeded for the first time to purify interferon to homogeneity out of tissue culture fluid by the use of this antibody. Alan Williams purified a transplantation antigen out of a cell lysate using a monoclonal antibody. These are only a few examples of how monoclonal antibodies suddenly opened new approaches to a host of so far untractable problems in biology and medicine. All these smashing successes had been reported within the first five years after the invention and by 1980 it was clear to everyone in the field that the hybridoma technology had generated a revolution in biomedicine. Moreover, what had been achieved by 1980 with monoclonal antibodies was only a beginning, many more breakthroughs could be anticipated.

There are numerous science prizes and the prize committees have to suggest new prizeworthy candidates every 1–2 years. As a consequence, the committees are notoriously under pressure to identify outstanding scientific discoveries, preferably of the type that is not outdated, obsolete, or proven wrong, within a short while. This is not an easy task as real and lasting breakthoughs in science are rare, and thus the invention of the hybridoma technology was the preferred target for science prizes in the years around 1980. In the beginning of the 1980s it became more and more obvious that monoclonal antibodies would be among the future subjects under consideration for the Nobel Prize in Physiology or Medicine. As head of the laboratory it was initially Milstein alone who had collected several major science prizes for the invention of monoclonal antibodies. Prize committees frequently assume that the postdoctoral fellow doing the experiments with his/her own hands at the bench is merely executing the ideas of the senior scientist. While this may sometimes be correct, in the case of monoclonal antibodies it was clearly not the case (see Chapter 6). Fritz Melchers had in the meantime followed Niels Jerne as director of the Basel Institute of Immunology and it was his board of trustees that alerted him to the unpleasant possibility that Milstein may eventually receive the Nobel Prize without Köhler. The Basel board of trustees included some prominent names in science, for example Nobel Laureates Renato Dulbecco, Jim Watson, and Manfred Eigen, who were among those who are asked by the Nobel committee for suggestions. According to Melchers, these members of the board decided to start a concerted action by soliciting multiple prominent scientists to suggest Köhler, in addition to Milstein, for the Nobel Prize. The campaign was finally successful and Köhler shared the 1984 Nobel Prize in Physiology or Medicine with Milstein and Jerne. Melchers: "The only thing we were surprised about was that Niels was included. Nobody had expected that."

Köhler lived with his family in a small German village, Efringen-Kirchen, a short distance north of Basel. Among the many celebrations and receptions for Köhler that followed the announcement of the Nobel Prize, Melchers recalls a reception by the mayor of Efringen-Kirchen as particularly touching. In his speech the mayor told the story how he was alerted to the fact that one of his villagers had been awarded the Nobel Prize. On the particular Monday morning his secretary came to him saying that a German from Munich had received the Nobel Prize. While he was only slightly interested in the news, his attention rose when the secretary came again to announce that the person was born in Munich but worked in closeby Basel. Finally all bells began to ring after another short while when it became known that the recipient of the prize worked in Basel but lived, of all places, in Efringen-Kirchen. The people of Efringen-Kirchen, mostly owners of small farms and vineyards, had known and accepted Köhler and his family as friendly but somewhat unusual outsiders. With the Nobel Prize he became the celebrity of the village and the farmers brought specimens of the produce of their farms to the reception as gifts for Köhler, much like at a Thanksgiving ritual. Melchers was also among the colleagues invited by Köhler to attend the Nobel ceremony in Stockholm. He reports that it went according to the established rites, exactly as described by Carl Djerassi in "Cantor's Dilemma". The only noteworthy exceptions were the lectures the Nobel laureates had to give at the Karolinska Institute the following day. Niels Jerne, who liked wine, was so drunk that he could hardly make sense out of his perfectly prepared manuscript. Köhler, who was second, was unexperienced enough to have planned his lecture with far too many slides. When he ran out of time, he started to rush through his slides, loosing his audience in due course. Milstein was last and and rescued the event doing his job well.

Figures 16, 17, 18
Pictures taken at the Nobel Prize ceremony. Köhler with Niels Jerne, Köhler with Cesar Milstein, Köhler receiving document from His Majesty King Carl XVI Gustaf of Sweden. (KEYSTONE/Pressens)

# Chapter 8
# The patent disaster

The scandal surrounding the monoclonal antibody patents marked a period of drastic change in the relationships between academic and commercial biomedical sciences. Before that period, academic biomedical research rarely had resulted in products of commercial interest. Commercial aspects such as patent protection were thus not prominent in the minds of academic researchers, not to speak of spin-off companies and the like. Conversely, pharmacological research was to a large extent based on random screening of large collections of chemicals or biologicals. Most drugs of therapeutic use had been identified by this approach, which in turn was of little interest to academic science. Although there had been remarkable exceptions, the commercial and academic worlds existed as separate entities. All of this drastically changed in the 1970s with the advent of two novel sets of technologies, molecular biology and monoclonal antibodies. Molecular biology, the science that suddenly permitted to play with genes, led to gene technology, the science that allowed the isolation and mass production of essentially any protein of human origin in bacteria. Products such as insulin, that had to be isolated from pig pancreas before, could now be produced in industrial fermenters, less expensive, more reliably, and in better quality. Research in molecular biology promised many more useful products to be developed by gene technology. An unprecedented goldrush set in, scientists started to think in industrial terms, venture capital became available, biotech companies were founded by the hundreds, patent protection all of a sudden became an issue, and stock options became a factor in the life of university professors. It was a revolution in science, technology transfer, and academic mentality. All of this took place primarily in the US, while in Europe, in Germany more than in other countries, authorities spent their time with lengthy considerations of biosafety and morale, resulting in inhibitory legislature and exodus of the pharma industries to the US. The hybridoma technology was part of this development. Up to now, procedures involving antibodies used polyclonal antisera produced in animals which had two drawbacks, every lot produced was different and of limited quantity no matter how many animals had been immunized. Standardization was a neverending issue. As a result, commercial products on the basis of antibodies were not, at a large scale, attractive to pharma companies and their market was rather limited.

There was also a big change in thinking about the issue of free access to scientific information as opposed to intellectual property. Around the time when Köhler and Milstein were awarded the Nobel Prize the notion was widely held that they had purposely and consciously refrained from patenting their hybridoma technique. This view was propagated for many years even in learned journals, such as *Scrip Magazine*, published in Britain and among the most respected international periodicals in pharmaceuticals, that commented in 1993: "*It seems incredible that in the mid-1970s two British scientists who discovered how to make monoclonal antibodies decided not to patent their invention. In accordance with long standing scientific tradition they felt they had no right to try to benefit commercially and personally...*" On that basis, Köhler and Milstein were frequently praised for holding the idealistic and altruistic spirit that the technique should be freely available so that its application by the scientific community would not be constrained by commercial interests. All of this is a misconception – Milstein himself had approached the authorities in charge with the aim to protect the hybridoma technique by a patent, but the attempt failed. Today the basic hybridoma technique is largely unprotected, it is widely used and commercially exploited by many without paying license fees. Until 1986 the number of patents relating to the hybridoma technique was estimated to be about 830, a small number compared to the explosion of its commercial exploitation in those years. Only a few prominent lawsuits have taken place (see below) presumably because the technique is being performed in so many variations that many patents are unlikely to withstand the legal test in court. While Milstein's initial attempt at patenting was unsuccessful in Britain, American scientists not much later obtained US patents on special applications of the hybridoma technique. Because the financial implications were estimated to be tremendous, the issue developed to what became a steadfast scandal in British science politics.

For several years after publication of the original paper by Köhler and Milstein (August 1975, *Nature*) there was very little public discussion on the pros and cons of patenting the hybridoma technique. The scandal started in 1978 when Hilary Koprowsky and Carlo Croce, both at the Wistar Institute in Philadelphia, filed two patents covering the hybridoma technique for two very broad applications, for cancer and for viruses. Koprowsky, Polish immigrant and scientific director of the Wistar Institute, and Croce, senior scientist at Wistar, had produced monoclonal antibodies to antigens on melanoma and on colon carcinoma cells, and these antibodies were considered for diagnostic and therapeutic applications in patients suffering from these malignancies. In addition, they were interested in antiviral vaccination and in the viral origin of cancer and had produced monoclonal antibodies against viral gene products that were of obvious interest for diagnostic use in virological laboratories. The experimental generation of these monoclonal antibodies at the Wistar Institute included use of Milstein's and Köhler's HAT-sensitive myeloma line. Milstein generously provided this cell line to anybody interested but, as a result of his failed attempt to obtain a patent, usually asked recipients of the cell line to sign an agreement in which they had to confirm that no commercial applications were pursued. Koprowsky's and Croce's patent applications were clearly a violation of this agreement, although they claimed not to have signed it. The two patents were issued in the US in 1979 and 1980, the US patent office apparently following the argument of the Wistar scientists that cancer or virus antigens were sufficiently distinct from the sheep red blood cells that Köhler and Milstein had used. The latter were highly immunogenic and therefore particularly suitable for the development of the technique, but of no practical value. Conversely, monoclonal antibodies to cancer or viral antigens were of pivotal medical interest while being much more difficult to generate. The patents were thereafter licensed to a US biotechnology company termed Centocor, in which the Wistar scientists held significant parts of the shares.

Koprowsky and Croce also submitted their patents to the British authorities. Predictably, the British office did not follow their arguments and rejected the application as an "obvious" application of the Köhler and Milstein technique, the technique was so-called "prior art". The US patents issued to Koprowsky and Croce had been deeply disturbing to British authorities who felt that the invention belonged to Britain and not to the US. As a consequence, a highly controversial public debate developed, characterized by bitter polemic. The filing of patent applications by American scientists in Britain on a technique developed under the leadership of a British scientist in a British laboratory was obviously seen by British politicians as an impertinence. The matter raised unpleasant memories of the penicillin affair of the 1930s. The British scientist Sir Alexander Fleming was the first to discover penicillin, first antibiotic ever observed, and had published the work without prior attempts at patenting. The situation seemed an exact precedent to monoclonal antibodies as Fleming had received the Nobel Prize for his pathbreaking discovery, while no patent was obtained on penicillin in Britain. Instead, American companies and scientists obtained the important patents on penicillin that resulted in billions of dollars in revenues. Since penicillin as a substance had been described and was thus "prior art", the American patents concerned the mass production of penicillin, an aspect that British scientists had neglected to pursue. Was a similar catastrophe now going to happen again with monoclonal antibodies? Were British scientists and authorities incapable of realizing the commercial perspectives of important discoveries? Were they unable to see the national advantage that was associated with products of such outstanding benefit to mankind?

British Prime Minister Margaret Thatcher was among those who held the popular notion that Milstein had, out of improper generosity, purposely abstained from trying to patent the invention. She was openly critical of

Milstein and on her initiative the British Research Councils and the Royal Society jointly appointed a committee to review the matter. Alfred Spinks, research director of the international pharma and chemical concern ICI, was appointed chairman of the Advisory Council of Applied Research and Development (ACARD) which delivered what became known as the Spinks Report in 1980, after two years of deliberations. Sections in this report accused Britsh scientists of negligence and irresponsibility, with unconcealed aim at the inventors of monoclonal antibodies: "*There appears to be a lack of awareness in practice of the obligations on recipients of government money and of the rights of NRDC* (National Research and Development Cooperation, the agency in charge of commercial exploitation of inventions made in public research institutes). *This must be remedied. We are concerned that a lack of appreciation of NRDC particularly by young scientists, may continue to result in situations such as that which occurred of monoclonal antibodies where patent protection was not sought early enough and British advantage was reduced.*"

Before this report had appeared, Milstein did not get involved in the discussion of patenting. Indeed, he rather seems to have enjoyed the reputation of scientific generosity and unselfishness that resulted from the lack of patent protection of the hybridoma technology. The same applied to Köhler, who did not contradict pertinent remarks in the multiple speeches or articles given or written in his honor. I remember a laudatio that I gave as director of the Max-Planck Institute of Immunobiology in Freiburg at a reception for Köhler on the occasion of the Nobel Prize. My praise of Köhler's and Milstein's conscious refraining from attempts at commercial exploitation went uncorrected by Köhler, but earned me intense lecturing on the importance of patent protection from several representatives of the local biotech industries during the following cocktail hour. Milstein, at least, changed his attitude after the Spinks Report, and began to insist that he had seriously considered patenting and had taken practical steps in this direction. Indeed, in the context of discussions on the safety of genetic engineering Milstein had presented the monoclonal antibody experiments during an internal meeting in Oxford of the Medical Research Council (MRC), i.e., in the presence of administrative staff of MRC. The commercial perspective of the work was noted and Milstein was asked to provide a preprint of the paper for consideration. He did this on 10 July 1975, i.e., about 1 month before its publication. A few days later he received a letter thanking him for sharing the information and continuing: "*...reading your paper confirms my feeling that this approach to antibody synthesis has great commercial implications. It is most important that action is taken as soon as any exploitable idea gets to the stage where patent protection can be applied for. I have therefore taken the step – and I do hope that you will forgive me for acting without consulting you first – of drawing the attention of the National Research and Development Cooperation to your reprint* (a misterm, it was a page proof, i.e., a replica of the printed pages provided by the publisher to the authors for final corrections before publication). *It may be that they will contact you direct to discuss whether to take any formal steps.*" With the submission of the paper to the MRC headquarters, several weeks before publication, Milstein maintains that he had done his duty to the British authorities in regard to patent protection. This is certainly correct, because with a rapid decision it seems to have been possible to delay publication for another few weeks until a patent had been submitted.

What was the role of MRC headquarters in the process? In the letter to Milstein quoted above the impression is put forward that the NRDC had been immediately alerted to the paper and its imminent publication. Particularly the reference to the lack of prior consultation with Milstein gives a strong impression of urgency. However, the account of the NRDC (later called British Technology Group [BTG]) on the events differs critically from that letter: "*...Dr. Milstein sent the MRC a printer's proof of the article only a few weeks before it was published. A further 13 months elapsed before the MRC formally* (!) *drew the article to the attention of the NRDC, and apologized to having failed to communicate it prior to publication.*" The interpretation of this statement strongly depends on the term "formally". Was the NRDC perhaps informally alerted prior to publication while a formal communication took 13 months? Did the NRDC perhaps neglect to react to the informal alert and waited to be formally informed? Did the MRC rest assured that an informal communication was sufficient to incite action in the NRDC? The matter was discussed at length at a meeting organized by the Wellcome Trust and published in the series "Wellcome Witnesses to 20th Century Medicine", at which all concerned parties were present and presented their views on the patent disaster. The question as to who had the major blame could not be clarified although it seems that the officer at MRC headquarters who first obtained the preprint from Milstein failed to push the matter in a way he should have done. Although the person was a scientist and should have been able to understand the subject, all participants at the meeting nobly agreed that the name should not be disclosed.

At whatever point in time it was alerted to the paper by the MRC, NRDC did not react for over a year after publication. Then, in October 1976, they answered to the MRC: "*...although they also suggest that the cultures which they have developed ... could be valuable for medical and industrial use, I think this statement should be taken as a matter of long-term potential, rather than immediate application. It is certainly difficult for us to identify any immediate practical applications which could be pursued as a commercial venture, even assuming that publication had not already occurred. I would add the general field of genetic engineering is a particularly difficult area ... and it is not immediately obvious what patentable features are at present disclosed in the Nature paper. In summary, therefore, unless further work indicates a diagnostic procedure or industrial end product that we can protect, despite the disclosure in the Nature paper, we would not suggest taking any further action ourselves*".

The NRDC letter, written in1976, highlights a problem that European scientists have compared to Americans. In the US, a patent can be filed up to a year after publication, whereas in Europe patents have to be filed before publication. The genuine and primary interest of scientists is to rapidly publish novel results in a so-called original paper. Obtaining a patent may be of interest for a company but is of no immediate value for a scientist aiming at an academic career. For young scientists, who do the work at the bench in the laboratory but have mostly a limited employment of 2–3 years, the publication of their work is the only way to document their work performance for job applications and career advancement. For established scientists publications are the only way to maintain professional reputation. Reputation depends on continued demonstration of scientific productivity and is a prerequisite for the continuing grant support by research funds that are in most cases provided by the funding agencies for limited periods of time, mostly 2 or 3, rarely up to 5 years. Before a grant ends, a new application has to be written in which the successful conduction of the previous application has to be documented by publications. The procedure of publishing scientific results is well established but by far not straightforward, it is cumbersome and it may take considerable time from obtaining the results to seeing them published in a scientific journal. The first step, interpreting what the data mean and preparing a rough draft together with graphs and tables, is usually left to the young scientist who is the first author of the paper. The materials then go to the senior scientist directing the work and last author of the paper, who corrects the draft, making more or less drastic changes. The draft may be passed back and forth between the junior and senior scientist until both are satisfied. Very often there are additional contributors who have performed certain techniques or have provided materials and will also be authors, listed between the first and last authors in order of the importance of their contribution. Typically, an advanced draft is sent to these individuals for comments or corrections, a process which may take considerable time until everyone agrees to a final version. The choice of the journal to which the paper is then submitted depends on a number of aspects including the subject of the work as well as the quality, novelty, and general interest. Scientific journals are in a ranking order of reputation, determined by their impact factors that indicate how often papers published in a journal are cited. There are only a few high impact journals whereas many journals exist with intermediate and even more with low impact factors. A paper published in a high impact journal commands *per se* an advantage in prestige, independent of the actual quality of the research it reports. Many more papers are initially submitted to the high impact journals than they can publish. Many scientists have a tendency to overestimate the importance and quality of their own research and submit their papers to journals higher in the ranking order than would be adequate. In order to select the papers for publication from the host of submitted manuscripts, high impact journals employ a two-tier process. In the first round, the editors mainly look at the author's names and subject matters of submitted papers and reject those which they consider not to be appropriate. The international standing of the senior author plays an important role in this initial selection, often more than the scientific impact of the work. The small number of papers surviving this round will then be subjected to peer review. The paper is sent to two established scientists in the field who rarely share the author's enthusiasm for the work. They decide on rejection or acceptability and, in the latter case, mostly suggest more or less drastic changes which may include whole new sets of experiments. As a result, the vast majority of papers are rejected on first submission. In that case, or if they are unwilling to comply with the changes suggested by the referees, the authors submit the paper to the next journal down the ranking order and the process starts anew. All journals demand a statement from authors that the work is not under consideration for publication elsewhere, so that simultaneous submission of a paper to several journals, which would shorten the time to publication, is precluded. When all goes well a paper may be published in 6 months, but more often than not it may take a year or more. Delaying publication by patent applications, which may take another year or more in Europe, therefore is hard to bear for scientists whose professional existence depends on published papers.

The issue of patents on monoclonal antibodies did not end there, it went on. There were two prominent lawsuits in the following years on special applications of the hybridoma technology. One concerned a patent obtained in 1980 by the US start-up company Hybritech on the use of monoclonal antibodies in diagnostic assays, so-called sandwich immunoassays. In this type of diagnostic assay, two antibodies are utilized to detect and quantify substances of interest in body fluids, for example hormones indicating pregnancy. One of the antibodies is attached to a plastic surface. The body fluid is then layered on top so that the target substance can attach to the antibody. The body fluid is washed away and the substance is revealed by the second antibody which has been labelled to give a color reaction that can be measured in a spectrophotometer. The US company Abbot had several such assays on the market utilizing conventional polyclonal antibodies. The Hybritech patent attempted to cover the Abbot assays as well as all future sandwich immunoassays using monoclonal instead of polyclonal antibodies. With monoclonal antibodies not only could one establish many more sandwich immunoassays, they also were more reliable and more easily manufactured. In a first lawsuit involving another start-up, Monoclonal Antibodies Inc., the Hybritech patent was declared invalid as the technique described "was obvious and logical to anyone skilled in the field". Although there were a number of appeals, several courts maintained this judgement until Hybritech was taken over by the multinational pharma giant Eli Lilly who decided to take the case to court again in 1986. Now, to no great surprise, all previous judgements were reversed and the patent was declared fully valid. Moreover, the US Supreme Court ruled in 1987 that no further appeals would be heard in this matter. Thus, sandwich immunoassays using monoclonal antibodies are patent protected once and forever. The financial significance of this decision can only be underestimated. Thereafter, the companies owing license fees to Eli Lilly included Johnson & Johnson, Becton-Dickinson, Bristol-Myers, Eastman Kodak, Amersham and Abbot.

The second, perhaps more interesting case concerned patents filed in 1979 and granted in the early 1980s to Ortho Pharmaceutical Cooperation protecting a number of monoclonal antibodies to human T cell marker antigens. The antibodies had been produced by Stuart Schlossman, Ellis Reinherz and co-workers at the Dana Farber Cancer Research Center of Harvard Medical School, Boston. These antibodies were among the first examples demonstrating the use of monoclonals for differentiation of human T cell subtypes. Today there are hundreds of such monoclonals available, allowing the distinction of any type of blood cell including T and B cells as well as other leucocyte types and subtypes. The production of the first such monoclonals and the demonstration of their usefulness as typing reagents was a scientific sensation and had warranted publications in the highest impact journals. In parallel, a sophisticated new technology was being developed for application of monoclonal antibodies in research and clinical laboratories, termed flow cytometry. Conventional antibodies had been used for many decades as reagents to differentiate cells using the fluorescence microscope. The cells to be studied were exposed to antibodies that were conjugated with fluorecent dyes and inspected by eye under a microscope that detected fluorescent light. While this is a useful technique, it suffers from lack of objectivity. The person examining the sample must decide if a cell is fluorescent or not, a sometimes difficult decision as the signal brightness can vary from cell to cell, and every cell has a certain amount of autofluorescence, the so-called background. In flow cytometry fluorescence is detected by an instrument independent of the human eye, the fluorescence-activated cell sorter (FACS). The cells to be diagnosed are passed in a stream of droplets, each containing a single cell, by a laser excitation system. The light emitted by the fluorochrome is detected by a photocell and the information is integrated by a computer for display on a screen. A FACS can determine several different fluorescent lights in one step, allowing the simultaneous use of several different monoclonals, each labelled with a different fluorochrome, and thus a highly sophisticated subsetting of complex cell populations. In addition to analytical cell differentiation, flow cytometers were soon developed that permitted the preparative separation of cell mixtures.

Ortho was one of two competing companies in the initial development of flow cytometry. In combination with monoclonal antibodies flow cytometry promised to become a billion dollar market. Ortho got together with the Harvard scientists and together they patented their first eight monoclonal antibodies for differentiation of T cell subsets, the so-called OKT series. As a prerequisite for obtaining patents on monoclonal antibodies, the US patent authorities had in the meantime issued that the hybridoma cells producing the antibody be deposited at the American Type Culture Collection (ATCC), the public repository and distribution agency for biologicals at the NIH. For research purposes, every scientist should be to able obtain the hybridoma cells from ATCC for the cost of shipping, isolate the monoclonal antibodies and use them in the laboratory without paying license fees. Nevertheless, not much later Ortho lawyers wrote a letter to a number of scientists who had done so, warning them that they may have infringed Ortho's patents, possibly resulting in financial consequences. This was followed by considerable anxiety in the scientific community, not only because of the imminent license fees. If the Ortho's lawyer was right, accessibility of available monoclonals for research purposes would be considerably hampered in the future. NIH patent attorneys replied that the use of materials deposited at ATCC was free of license fees and did not infringe patents associated with these samples, as long as they were used solely for research purposes. Ortho did not pursue the matter and no legal action was taken in this case.

In the meantime, Becton-Dickinson also had generated a number of monoclonal antibodies for human T cell subsetting. More than Ortho, Becton-Dickinson had done most of the basic technical development in flow cytometry, working together with Lee and Len Herzenberg at Stanford University, Palo Alto. Together they had developed the first prototype FACS instruments that were simple enough to be used by ordinary scientists in their laboratories, and Becton-Dickinson has remained the leading company in this technology until today. Similar to Ortho, Becton-Dickinson wanted to engage in selling monoclonals for use with their flow cytometers. While they were independently generated and therefore undoubtedly structurally different, the Becton-Dickinson monoclonals were directed against the same T cell antigens as the OKT series of Ortho. Using this argument, Ortho filed a patent infringement suit against Becton-Dickinson in 1984. A legal argument ensued about whether the structure of a monoclonal or its functional specificity was the critical property that counted in patent litigation. The case was settled out of court, but strongly in favor of Ortho. Becton-Dickinson paid US dollar 5 million for royalties and license fees in return for nonexclusive licenses on the patents, excluding putative future therapeutic use of the monoclonals.

M. MacKenzie, P. Keating, and A. Cambrosio have given the patent issue of monoclonal antibodies a thorough treatment from a lawyer's viewpoint in 1990. They conclude with the following statement: "*We do not bemoan the naiveté of scientists, or take a moral umbrage at the cynicism of lawyers or the contamination of scientific practice by its commercial possibilities. Our conclusion is simply that from an analytical point of view there are significant changes taking place in the political economy of science and technology, with strong implications for the way science is done. To take an active role in those changes as a group, scientists may well have to reflect upon their professional activity.*"

# Part II

# The time after

# Chapter 9
# The Max-Planck-Institute of Immunobiology

In April 1984 Köhler accepted the offer to become a member of the Max-Planck Society for the Advancement of Science (MPG), associated with a directorship at the Max-Planck Institute of Immunobiology (MPIIB) in Freiburg, next to the Black Forest in the southwest of Germany. The MPG is a research organization unique to Germany and hard to compare with the science structures operating in other countries. A directorship in the MPG not only means generous resources for the rest of one's professional life, one is also accepted in a small circle of scientists, no more than about 250 at anyone time, who represent the top level of German science. The society originated from the Kaiser-Wilhelm-Gesellschaft (KWG) founded under the patronage of the German Emperor in 1911 in Berlin. Emperor Wilhelm II, who ruled over Germany until he was forced to retire at the end of World War I, wanted what he called a "German Oxford". The ministerial director in the Prussian Ministry of Culture and Education, Friedrich Althoff, was charged with making the initial plans for the establishment of a number of non-academic research institutes. Dahlem, then an idyllic and partly agricultural suburb in the south-west of Berlin, was chosen for a site, as there was no space available in the center of Berlin. For the same reason, Althoff had already moved a number of institutes belonging to Berlin's Friedrich-Wilhelm-University to Dahlem. Adolf von Harnack, director of the Imperial Library and later first President of the KWG, strongly endorsed Althoff's plans and wrote a memorandum to the emperor in which he compared the desparate situation of science in Prussia with that in France, England and the US. Following a narrow interpretation of Wilhelm von Humbold's principle of "unity of research and teaching", publicly financed research in Germany was almost entirely done in universities. The foundation of the KWG was announced on 11 October 1910, at the celebrations for the 100th anniversary of the Friedrich-Wilhelm-University, by the Emperor himself: "*Humbold's great scientific plan demands, along with the Academy of Sciences and the universities, independent research institutes as an integrating component for the entire scientific organism. ... We need institutions which exceed the boundaries of the universities and are unimpaired by the objectives of education but in close association with academia and universities, serving expressively the purpose of research. To be able to guarantee the facilities a permanent endowment, it is My desire to establish under My protectorate and in My name a society whose task it is to establish and maintain research institutions. I will gladly supply the society with the means at My disposal. It will be the task of My government to see to it that government aid is available to the extent that it is needed. May today not just mark the celebration of the Berlin University, but simultaneously be a further step in the development of the German intellectual spirit*".

The actual foundation of the KWG, today the Max-Planck-Society, took place on 11 January 1911. Its first institute was the Kaiser-Wilhelm-Institute of Physical Chemistry and Electrochemistry, founded in 1912. Its founding director was Fritz Haber, best known for inventing the technology to convert atmospheric nitrogen into ammonia. It was renamed the Fritz Haber Institute in 1956. As a bynote, it had already become obvious that the Emperor's financial support, so generously promised in his founding speech, was not at all sufficient. A private foundation had to be formed to make significant contributions so that the building could be completed. Also with considerable outside contributions, in this case from industry, the Kaiser-Wilhelm-Institute of Chemistry was the second foundation, in 1912. It consisted of three departments under the direction of Ernst Beckmann (physical chemistry), Richard Willstädter (organic chemistry), and Otto Hahn with Lise Meitner (radiochemistry). One year later followed the Kaiser-Wilhelm-Institute of Experimental Therapy, established at the initiative of Paul Ehrlich and his student, the immunologist August von Wassermann, who had invented the Wassermann reaction for the diagnosis of syphilis while working in Berlin's Robert-Koch-Institute. As founding director of the Kaiser-Wilhelm-Institute, Wassermann extended his studies on serodiagnosis of infectious diseases to trypanosomiasis and tuberculosis, with less success than for syphilis. He was soon joined by a co-director, the biochemist Carl Neuberg, who identified new therapeutic derivatives of arsenic, quinone, and salicylic acid.

Until 1930, ten additional Kaiser-Wilhelm-Institutes were established in Berlin Dahlem, together constituting the unique and top-level scientific establishment which soon became known in the world as the "Dahlem Domain". Among them were institutes as famous as the Institute for Cell Physiology with its director Otto Warburg, the Institute of Biology with Hans Spemann as vice-director, the Institute of Biochemistry which later became Adolf Butenandt's working place, as well as institutes of technical orientation such as the Institute of Fiber Chemistry and the Institute of Silicate Research. The unique assembly of outstanding scientists of international reputation created the legendary "Spirit of Dahlem". On the other hand, Dahlem also was home to Kaiser-Wilhelm-Institutes that later used pseudoscientific approaches for justifying the eugenic and racial policies of the Nazi regime, such as the Institute of Anthropology, Human Genetics, and Eugenics. With the forced exodus of Jewish scientists from Germany in the mid 1930s the spirit of Dahlem began to fade away. During World War II almost all Kaiser-Wilhelm-Institutes were relocated to other German cities and any new institutes had been established outside Berlin as well, for reasons of safety. During the chaos that followed the German capitulation the president of the KWG committed suicide and the general secretary, Dr. Telschow, moved the central offices of the KWG to Göttingen, single-handedly with the help of his three secretaries. There he persuaded Max Planck to take over the presidential responsibilities. Planck, who had been president from 1929–1937, used his impeccable reputation and negotiated with the allied military administrations about the future of the KWG. Together with the remaining number of KWG institute directors still available in 1946 he organized the election of Otto Hahn, then working in England, as new president. Hahn accepted and his first task was to avert a resolution of the Allied Control Council to dissolve the KWG. He succeeded with the help of a farsighted official of the British military administration, Colonel Dr. Bertie Kennedy Blount. Blount had studied in Germany, was fluent in German, and developed a close friendship with Telschow while being in charge of academic institutions in the British sector. He did not hesitate to issue manufactured permission papers to Telschow so that he was able to cross borders to the French, American, and Russian sectors in his travels visiting what was left of KWG institutes in various parts of Germany. Most importantly, Blount persuaded his superiors to support the continuation of the KWG under a new, non-offensive name. With the consent of the British military administration, the Max-Planck Society for the Advancement of Science was constituted on 11 September 1946, as a successor of the KWG, first in the British sector but later also in the American and French sectors. Otto Hahn was president and the central offices were officially installed in Göttingen. With the foundation of the Federal Republic of Germany in 1949 a state treaty with new bylaws was adopted for the MPG. Most importantly, the annual budget of the MPG is provided in equal parts by the federal budget and by the combined budgets of the Länder. Moreover, in spite of the public funding, the MPG is legally a private foundation and entirely independent in its decisionmaking.

In the following years of the "Wirtschaftswunder" (economical miracle) the MPG expanded to more than 65 institutes distributed throughout West Germany. The most dramatic expansion took place in the 1960s under the presidency of Adolf Butenandt who oversaw the acquisition of the Max-Planck-Institute of Immunobiology, as well as the foundation of many other institutes. While most of the new institutes in the MPG were established from scratch with new institute buildings, facilities and personnel, some were acquired, completely or partially, as existing structures previously belonging to universities or other research organisations. Most unusual was the acquisition of the Max-Planck Institute of Immunobiology (MPIIB) which had been an industrial institute before and was purchased by the MPG, including its director and personnel, for the sum of 3 million marks. According to the Harnack principle, named after the first president of the KWG and used as a guideline in its hiring policy until today, the MPG tries to identify outstanding scientists, independent of their field of research, and does whatever it takes to persuade such individuals to join the MPG. In the case of the MPIIB, the individual at stake was Otto Westphal.

Westphal was born in Berlin-Charlottenburg in 1913, a time which he himself referred to as the Belle Epoche of Science in Germany, borrowing the term from the arts. His parents had been part of the intellectual elite in Berlin. His father, Wilhelm Westphal, was a professor of physics and member of the Physics Faculty of the Friedrich-Wilhelm-University. It included scientists as prominent as Max Planck, Albert Einstein, Max von Laue, and Gustav Hertz, which were regular guests at the Westphal's home. Westphal later wrote "*I had no idea how privileged I was to share their company*". Westphal's mother, Olga, a gifted pianist, played four hands with Max Planck and accompanied Einstein who played the violin. She is quoted to have written in her diary "*I joined in the singing (of Brahms' Liebeslieder performed at her home by a choir directed by Max Planck) so intensely that Otto was born 10 days early…*" When the parents divorced in 1925 Olga Westphal moved to Freiburg with her four children, where Otto lived most of his life. At age 28 Otto married Dr. Olga von Gayling, heiress of the castle Schloss Ebnet in Freiburg and reportedly a spectacular woman of pronounced emancipation who had obtained her PhD in Physics and took part in car racings. They had two children who preferred to use the name von Gayling after Westphal left the family to live with Ursula, his long-term technical assistant and nearly 30 years his junior. Olga von Gayling became an alcoholic and had to be committed to a nursing facility. Several years later, Westphal divorced Olga and finally married Ursula in1976. She gave birth to a son in 1981, an event that earned Westphal similar admiration than his scientific achievements. After the family had moved to Montreux, Ursula developed symptoms of psychosis and in 1986 she committed suicide by jumping from a window. Referring to his relationships to women, Michel Sela once commented on his friend Westphal: "Otto is a monster". In 1987 Westphal married Denise Josette Mottier, a Jehova's Witness, who was his next door neighbor. She took care of him until he died in 2004 at 91 years of age.

Westphal studied chemistry, the leading science at the time, and became fascinated by the concepts ot stereochemistry introduced by Jacobus van't Hoff, Emil Fischer, Paul Ehrlich, and Svante Arrhenius. He approached Karl Freudenberg, pupil of Emil Fischer and professor of Chemistry in Heidelberg, for a thesis and was given the task to determine the chemical basis of the blood groups, discovered many years earlier by Karl Landsteiner. In 1937 he obtained his PhD and thereafter worked as scientific assistant in the Kaiser-Wilhelm-Institute of Medical Research, in Heidelberg, on the synthesis of vitamins that had previously been discovered by the director of the institute, Richard Kuhn. During this time he continued his studies of immunology and serology by listening to the lectures of Hans Sachs, a pupil of Paul Ehrlich and professor for experimental cancer research at Heidelberg University before he had to leave Germany and moved to England. Sachs was interested in serodiagnostic approaches to cancer and eminent immunologists like Ernst Witebsky and Erwin Neter, who later emigrated to Canada, had received their initial training in his laboratory. As a chemist with experience in immunology,Westphal was one of the first in Germany to understand the modern concepts of immunochemistry. At age 29 he was appointed head of the department of biochemistry in the institute of chemistry in the medical faculty of the University of Göttingen. He was in charge of teaching biochemistry to the medical students, a task that saved him from being drafted to the German army. Here he met Otto Lüderitz and Botho Kickhöfen, who stayed with Westphal for their entire scientific careers. The group collaborated with F. Schütz, professor of hygiene in Göttingen, who was working on bacterial vaccines. Together they extracted *Proteus* OX19, that cross-reacted with *S. typhi* (known as the Weill-Felix reaction) and seemed a promising typhus vaccine candidate. A laboratory worker got contaminated and developed a high but transient fever, an incident leading to the identification of the first purified bacterial pyrogen. Fever therapy was an important medical regime at the time and the discovery soon made Westphal well known in the field. While he stayed with the subject ever since, he changed to the pyrogen lipopolysaccaride (LPS), initially described as endotoxin by Richard Pfeiffer in 1892. Westphal started to work on LPS in 1947 after he had accepted an offer of the Swiss pharmaceutical company Dr.Wander AG, best known as manufacturers of Ovaltine, to head a new institute in Bad Säckingen, near the Swiss border. This and the Swiss ownership of his institute gave Westphal research opportunities and international connections that few other German scientists were able to enjoy right after World War II. Although Dr. Wander had hired him to produce antibiotics, Westphal was able to interest Wander in pyrogens and the group started to work on LPS. Clinical trials were done in collaboration with Ludwig Heilmeyer, then chairman of internal medicine in Freiburg. As a result, Westphal and his group developed highly purified Salmonella LPS, produced by Wander under the tradename Pyrexal, for human fever therapy.

Feeling that Bad Säckingen was a somewhat isolated place for an international research institute, Wander and Westphal agreed in 1956 to move the institute to Freiburg, near a major University. Westphal had become an internationally recognized scientist and started collecting prices, honorary memberships and medals from various national and international organizations. It was only natural that the MPG was alerted to this rising star on the still rather dark sky of German biomedical science. In 1961, a period of happy expansion of public spending in Germany, the decision was made to incorporate the Wander institute, including Westphal and most of the personnel, into the MPG. It is not known what made Dr. Wander decide to let the institute go for a rather symbolic price, it may have been pure philanthropy. It was named Max-Planck-Institut für Immunbiologie (MPIIB) and Westphal was appointed founding director. Later he was joined by two co-directors, Otto Lüderitz and Herbert Fischer. Among the most significant achievements in the 20 years of Westphal's directorship of the MPIIB was the demonstration that the biological activity of LPS was associated with the Lipid A moiety, whereas variations in the polysaccharide side chains were responsible for the serological specificities of the Kauffmann-White scheme that distinguished the various *Salmonella* substrains. Moreover, under the direction of Lüderitz, MPIIB scientists determined the structure of *E. coli* Lipid A, except for a few small details that were later corrected, using more advanced NMR technology, by the group of Tetsuo Shiba in Japan. For example, and somewhat sadly, MPIIB scientists had misplaced a fatty acid side chain on the Lipid A core. Synthetic products according to this structure failed to show the biological activity of isolated natural Lipid A in animals. Only after correcting this mistake, synthetic Lipid A had full biological activity, so that the final solution of the structure of LPS came from Japan rather than from Westphal's MPIIB.

No other scientist impersonates post-World War II development of immunology in Germany like Otto Westphal. Under his direction, the MPIIB became a nationally and internationally leading center of immunological research, and a breeding ground for generations of biomedical investigators thereafter holding key positions in research and medicine in Germany. He was the main initiator in the foundation of the Society of Immunology in Germany (DGfI) in 1967, which took place at a meeting in the Hoechst Hall of Centuries with most prominent immunologists of the time in Germany attending, among others Klaus Rajewsky, Gerhard Schwick, Norbert Hilschmann, Paul Klein, and Niels Jerne. Needless to say, Westphal was elected first president and held this office for 10 years. He played a key role in the foundation of the *European Journal of Immunology* in 1971, the MPIIB having been home to its editorial office ever since. In the Max-Planck-Society and in the Deutsche Forschungsgemeinschaft he chaired at one time or another, nearly all important decisionmaking bodies and committees. He was a member of multiple learned academies in Germany and abroad, and received numerous honors and prizes including, among many others, the Emil von Behring Prize (1964), the Paul Ehrlich Prize (1968) and Robert Koch Medal (1983).

Westphal's charismatic personality held many facettes beyond that of a prominent scientist. Well known was his cultural aura. He was a skilled player of the flute and loved to organize classical music recitals at his institute where he performed together with professional musicians. Well remembered are the symposia termed "Reflexions" where Westphal invited his friends to listen to music and discuss the deeper meanings of life. Regular guests were eminent scientists including Sir Peter Medawar and his wife Lady Jean, Lewis Thomas, Manfred Eigen, and outstanding artists such as the cembalists, Fritz Neumeyer, collector of and performer on historical keyboard instruments, and Edith Picht-Axenfeld. Sara Sela once said to me: "Otto is a very ceremonial man. But is there much else in life?" There was also the godfather facette. If one had to find a position or had some other career related problem, one went to see Westphal. When the German military service sent me a draft notice in the middle of a postdoc period at Rockefeller University in New York, it was Westphal's influence that saved me from having to follow. He paved the career path for many of us, independent of whether one had trained in the MPIIB or elsewhere. There was also his juvenile membership in the Nazi SA, a fault that burdened him his entire life. I was blamed, as chairman of the VII International Congress of Immunology in 1989 in Berlin, for having invited Westphal to serve, together with Niels Jerne, as Honorary President. The most outspoken colleague in this affair, James Dinarello, wrote a critical and accusing letter to an innocent German postdoc applicant who in full shock passed the reply on to me. Dinarello then agreed to meet Otto Lüderitz, who knew first hand about Westphal's attitude towards the Nazi regime. After the two had a walk in the Black Forest, the affair came to an end, although Dinarello could not be convinced by Lüderitz. For my own attitude towards Westphal it was significant that many Israeli scientists were in friendly contact with Westphal, who had been a driving force in the early post-war relationships between the Weizmann Institute of Science and the Max-Planck Society. Michael Sela of the Weizmann Institute of Science became a close personal friend of Westphal and accepted in the early 1970s to become an external scientific member at the MPIIB.

At the celebrations for the 40th anniversary of the MPIIB in 2001, Westphal met the late Charles Janeway Jr who was among the invited symposium speakers and who had begun his lecture with the remark: "I should particularly like to thank Otto Westphal who spent years to purify my favorite molecule, which is LPS." I had the privilege to witness this unique encounter between these two outstanding scientists, Westphal the chemist who elucidated the structure of the molecule and Janeway the biologist who defined its crucial role in immunity nearly half a century later. During the meeting Janeway said that he was happy to meet Westphal for the first time and the two agreed that it would presumably also be the last. At this occasion, 20 years after his retirement, Westphal gave a lucid speech commemorating the foundation of the institute, and was honored by standing ovations. He was clearly a remarkable man.

Until the time of Westphal's retirement in 1981 the MPIIB had made only halfhearted efforts to include studies on the immune system of the vertebrate host, i.e., to fulfil the legacy of its name, immunobiology. As a step in this direction, Herbert Fischer had been brought in as a co-director.

However, in keeping with the main avenues of thinking at the insitute, his interest centered on non-specific immunity, particularly macrophages and their mode of activation. Most of the work in the institute still focussed on bacteria, and the institute's respectable animal facility largely served to test parameters of unspecific immunity, such as the rejection of tumors by LPS, later shown by others to be due to the cytokine tumor necrosis factor (TNF). In addition, MPIIB scientists had synthesized a novel class of lipids, termed alkylphospholipids, that showed anti-tumor and immunosuppressive effects. Only a few scattered projects on specific immune mechanisms, i.e., antibodies or T lymphocytes, were pursued by individual MPIIB scientists. To the scientific community in Germany and abroad research at the MPIIB did not seem very modern at the time and the impression prevailed that the institute had not kept pace with the rapid developments in molecular and cellular immunology that were going on elsewhere. The Basel Institute of Immunology was in its first decade and its director, Niels Jerne, had convinced most of the scientific community that the one pressing problem left to be solved in immunology was the genetic basis of immunological specificity. Within a few years, the international reputation of the MPIIB was clearly outgrown by that of its new neighbor, the Basel Institute. Moreover, Klaus Rajewsky's Department of Immunology at the Institute of Genetics in Cologne with its pathbreaking contributions on cellular cooperation in the immune system began to draw much international attention. It was therefore strongly felt in the German scientific community that a young molecular or cellular immunologist should be selected to succeed Westphal. I was a junior group leader in Rajewsky's department at the time, and remember Rajewsky joking: "We have to take over Freiburg". The MPG, however, has its own procedures for selecting candidates which expressedly do not include outside suggestions, not to speak of outside pressure. On the contrary, the existing institute directors, officially – but only officially – excluding the director to be replaced, first identify who they think might be the ideal candidate and suggest that person to the president of the MPG for consideration. The candidate is then evaluated by a committee consisting of other MPG institute directors of more or less related fields. The committee is free to co-opt outside scientists as committee members, if it feels that its own competence is insufficient. Outside scientists co-opted or not, the most important elements in the decision of the committee are written reviews solicited from six to eight international experts. If positive, the recommendation of the committee is then voted upon by the assembly of directors in the section of the MPG to which the institute belongs (in this case biology and medicine), and finally adopted as a resolution by the senate, the highest governing body of the MPG. Thereafter, the president starts negotiations with the candidate.

The first candidate for the Westphal succession favored by the institute was Lars Ake Hanson from Gothenburg, a clinical immunologist and pediatrician. Then in his early forties, he had received postdoctoral training with

Ouchterlony in Sweden followed by Grabar in Paris and Kunkel in New York. He had earned international recognition by his discovery of secretory IgA and further studies on its secretion at mucosal surfaces. His present field of study was the role of the mucosa in the defence against bacterial infections, with a focus on Gram-negative bacteria and their antigens. With the bacterial orientation of his research he fitted well into the traditional scope of the MPIIB. At the time he was head of the Department of Clinical Immunology and Microbiology of the University of Gothenburg, and he could not be persuaded to even consider leaving his present position. As a result of his being suggested for a directorship of the MPIIB, the University of Gothenburg had promoted him to physician-in-chief of the university's pediatric hospital, in addition to the post he already held. Thus, the search procedure in Freiburg had to be started anew and the institute centered on Rajewsky who had already been discussed in parallel with Hanson. Rajewsky was famous at the time for his work on the role of T helper cells in specific immune responses, but was certainly not in keeping with the institute's focus on bacterial immunochemistry. Nevertheless, he had become the most prominent German immunologist of the young generation, was the son of a former Max Planck director, and probably was acceptable to Westphal because of his outstanding command of the piano. Indeed, phantasies have been reported on placing a grand piano in the lecture hall of the MPIIB for Rajewsky to play at his leasure. However, in spite of his jokes about "taking over Freiburg", he turned it down after a lengthy period of negotiations. While Rajewky may have had other, perhaps more personal, reasons for staying in Cologne, the general interpretation was that the new directorship at the MPIIB was not very attractive because all positions, scientists and technicians, were filled with permanent personnel.

After this double fault the president of the MPG, Reimar Lüst, became irritated with the MPIIB, as it was highly unusual that offers of directorial appointments in the MPG were turned down. As a result, Lüst informed the MPIIB that he would not accept further suggestions by the institute until a number of free positions were provided to be included in the resources offered to the new candidate. The other remaining candidates in Germany were Fritz Melchers and myself. Both of us were interested in the specific branches of the immune system, but the MPIIB's resistance against leaving non-specific immunity and bacterial immunochemistry had broken down already with Rajewsky. Melchers was working in the Basel Institute (see Chapter 3) and his recent work on the diversity of B cells by limiting dilution analysis had raised considerable international interest. I was head of the small Unit of Cellular Immunology in the German Cancer Research Center in Heidelberg since 1976, after having worked in Rajewsky's department for nearly 5 years. I was interested in understanding specific antigen recognition of T cells and in the idiotypic network that, according to Jerne, regulated the immune response. Both of these were topics of considerable international attention at the time. Westphal favoured Melchers whom he provided with purified LPS that Melchers used to induce antibody secretion of B cells in his limiting dilution assays. Melchers had already been considered in parallel with Hanson but his problem was his father, Max Planck director Georg Melchers, who was notorious in the MPG for his endless and controversial elaborations in section meetings, irritating the members by extending the time of many a meeting much beyond the anticipated end. There was a widespread fear that Fritz resembled his father not only externally, which he does, but also in his attitudes and behavior. I had been previously considered in parallel with Rajewsky and a minority of members of the committee that was evaluating Rajewsky had supported my nomination instead of Rajewky's. In the end, the decision was made in my favor. Westphal was not happy with the decision and wrote long letters of explanation to Melchers, of which I received copies.

When I arrived in Freiburg in early 1981 my expectation was that I had at least 5–6 years to concentrate solely on the building of my own laboratory. Westphal was still present at the institute and conducted his "Reflections" symposia, to the outside world seemingly still in full command. Herbert Fischer, in his early 60s, was acting director and expected to take care of the administration of the institute for the coming 3 years. Thereafter Otto Lüderitz, a few years his junior, was to become acting director for another 3 years. Only then I was to take over and engage myself with planning the future of the entire institute by selecting their successors. However, Herbert Fischer died suddenly of a heart attack only a few months after my arrival and, sad as this was, the opportunity arose to bring in a second new director and thus move the institute to the modern world of immunology much more rapidly than had seemed originally possible. On the other hand, hiring a director of a Max-Planck Institute can be a considerable hassle, taking much of one's time and energy for a considerable period of time. Not being fully aware of the latter, I decided to take over. I thus was appointed acting director of the MPIIB by the president of the MPG on 1 January 1982, my main duty being to reorganize the institute in order to set the stage for a scientific reorientation, in combination with the appointment of a new director. As a lucky coincidence, Westphal moved to Heidelberg as interim chairman of the German Cancer Research Center which badly needed a helping hand after its former chairman, Hans Neurath, had quit the job because of frustration with its invincible bureaucracy. Otto Lüderitz, when I did not accept his well meant offer to step in as acting director after Fischer's death, was back in his little office thinking about the chemistry of LPS. All of a sudden I found myself in full charge of the MPIIB.

Max Planck institutes typically consist of several departments, each under a director. The acting directorship rotates among the directors, with a 3-year term. The duties of the acting director are, in addition to the day-by-day administration, to act on behalf of the institute by negotiating with the president of the MPG, with the central administration in Munich, and in the assembly of the section, roughly similar to a parliament. The hiring of a new director, in combination with a new scientific orientation, requires elaborate preparations within and without the institute. Firstly, the acting director must coordinate the sometimes quite diverse ideas of the co-directors in the institute, as well as that of other senior scientists who may have acquired some degree of influence inside and outside the institute. Once a consensus has been reached in support of a candidate and field of research, the candidate can be proposed to the authorities in the MPG. Secondly, as a general rule the institute has to provide the resources for the new director by itself. A typical department has 5–6 position slots for scientists that can be filled at any level, from postdoc to senior scientist, which are given a corresponding degree of scientific independence. The choice of scientific co-workers is essentially left to the discretion of the director, but directors are obliged to plan ahead so that by the time of their retirement most, if not all of the scientist positions in their departments are unoccupied. The incoming director inherits the budget as well as the position slots of the department of the retired director. At the MPIIB, the situation was entirely different. Westphal and his co-directors had refused to establish departments. Instead, the approximately 30 scientists worked in small laboratory groups that had spontaneously formed by interest and mutual liking. In addition, many of the scientists at the institute were rather senior and most had tenured positions. As a result, there were virtually no resources free for a new director. The first problem to be solved was thus the generation of free positions, i.e., to persuade scientists to apply for positions elsewhere and to accept outside job offers if one came along. After about a year and a lot of discussions and phone calls, a number of free positions were in sight and the process of identifying a new director could begin.

## Chapter 10

# Getting Köhler to Freiburg

I had met Georges and Claudia Köhler first in 1973, at a joint meeting of the French and German societies of immunology in Strasbourg. Köhler was in the middle of his PhD work in Basel and had got to know some of my students, who were also working on problems related to antibody diversity, during the conference. We met with a crowd of other young immunologists in some pub in the Petite France quarter for supper and drinks. Georges was a tall fellow, about 190 cm, with a slim but broad and sportive frame. He impressed me and others mostly because his hair was essentially white in spite of his youth, and he wore that strange curly and grayish beard, which he let grow seemingly ungroomed, not trimmed in any way. He looked at you through his steel-rimmed glasses in a friendly way, seemingly amused with the variety of people that one comes across. He seemed very sure of himself, a property that distinguished him from most other students who frequently are self-conscious and inhibited in the presence of more senior scientists. On the other hand he was somehow withdrawn, not really interested in getting to know people. He spoke slowly and with a low voice which he rarely raised. He was quite a relaxed fellow then and had as much wine as everybody else but laughed rarely and did not participate much in the conversation. Instead, he left the communication to his wife, Claudia, who was rather charming and talented in this respect. Nevertheless, Köhler's presence was not to be ignored; he seemed to form the center of the party. In some way, it was my first impression that Georges was an unusual person, but I could not tell in which way.

We met off and on since that first encounter and developed a relationship typical of colleagues in science, a non-committal sympathy but not a friendship. From the late 1970s, with the rising interest in monoclonal antibodies, Köhler was on his way to international recognition and was frequently invited as a main speaker at conferences, where we ran into each other more and more often. Like many other colleagues, I was impressed by Köhler's past achievements and his ongoing work in Basel, and saw him on his way into the leading group of German immunologists. I was head of one of four laboratory units in the new Institute of Genetics at the German Cancer Research Center at the time. The fourth unit was still open and I suggested to Georges to apply. He indeed did so and was invited for a seminar and an interview by the search committee. The expectations in monoclonal antibodies as "magic bullets" against cancer flew high at the time, and Georges was asked if his plans were to engage in such research. His answer was a clear "no" and so he was not further considered for the position. I was positively impressed by his uncompromising answer and disappointed that the appointment failed. Like many other scientists, Georges and I almost never talked about scientific topics. Scientists working on different subjects, like Georges and myself, mostly have limited interest in the details of each other's work. On the other hand, scientists who work in the same field are competitors and therefore rarely inform one another of their unpublished results. Rather, when we talked it was the usual gossip about other colleagues and mutual aquaintances, but Georges was not the chatting type and therefore our conversations were mostly short. Nevertheless, I formed an impression of Georges as a person who was sure of his own competence as a scientist but introverted, not interested in public exposure, his mind primarily focussed on his work in the laboratory. This impression was a grave misjudgement but later became an important element in my decision to suggest Georges as my co-director at the MPIIB.

The idea that Köhler could be the right candidate came to me in the fall of 1982 when I unintentionally overheard a remark that the late Peterhans Hofschneider, then director at the Max-Planck-Institute of Biochemistry in Munich and at the time chairman of the biomedical section assembly, made in a coffee break conversation with another person: "we should get Georg (!) Köhler into the MPG". Hofschneider pronounced Köhler's first name in the German way, which would have made Köhler furious who strongly insisted on the French spelling and pronunciation, Georges. Hofschneider's remark prompted me to seriously consider Köhler as a co-director. He was an immunologist and thus a natural choice for the MPIIB. His work on B cells seemed a perfect complementation to the interest of my own laboratory in T cells. He was trained in molecular biology, forseeably the main approach of the coming years in the biomedical field. He seemed a reasonable fellow to get along with, 7 years my junior, still actively working in the laboratory and thus presumably willing to let me keep the lead in the development and outside representation of the institute. But would I be able to push him through the evaluation process of the MPG? He was rather young, 36 in 1982, a borderline age for a director in the MPG but not impossible. He had about 30 publications, a rather short list, but again not prohibitive. One of the questions that is always asked in the evaluation committees is: "what has the candidate done?" and as proposer one has to be able to give a short and convincing answer. I did not expect difficulties with answering that question for Köhler, who had invented monoclonal antibodies.

There was a fair degree of resistance from inside and outside the institute against Köhler. The efforts to provide free positions had selectively reduced the number of MPIIB scientists working on specific immune mechanisms. While they had taken the opportunities to leave, most of the traditional bacteriologists had stayed. Helen Mäkelä, senior microbiologist from Finland, was chairperson of the scientific advisory board and excerted her influence to make the institute stay on the beaten path. She formed a coalition with the fraction of bacteriologically-oriented scientists at the institute making suggestions of candidates with a bacteriological background. Candidates were invited to give seminars and their interest in joining the institute was examined. Whatever position a scientist occupies in Germany, no one ever says no to the possibility of an offer from the MPG. Some of the names put forward were quite strong and it was hard to argue against them. After endless discussions it became clear to me that the bacteriologists could not be convinced and that there was little support for Köhler to be expected from this group, still forming a majority at the MPIIB. If I were to put forward Köhler, it was in violation of the deeply felt sentiments of the MPIIB scientists and I would have to bear the consequences, however they would turn out. I was not happy with the situation and would have loved to propose a consensus candidate, but I became increasingly convinced during this period that Köhler was the right person with whom the institute could be developed into a modern center of immunological research. In the Max Planck Society there is no formal requirement that a new directorial candidate has the support of the institute's scientists, except for that of the directors. Since Lüderitz kept himself out of these discussions, it was on me to make the decision, and I decided to suggest Köhler.

On 6 July 1982 I wrote a letter to the president of the MPG, Reimar Lüst, suggesting Georges Köhler to succeed Herbert Fischer as third director of the MPIIB. In the end, much of the resistance of the scientists in the institute had worn out, and I could sell the suggestion as having the support of the entire institute. To give extra strength to such a proposal I had learned that it has proven helpful to embed the candidate into a novel concept for the future of the institute, thus demonstrating that one had given the matter far reaching considerations. My novel concept for the MPIIB consisted of the establishment of three departments, Mechanisms of Pathogenicity (director: O. Lüderitz), Cellular Immunology (director: K. Eichmann) and Molecular Immunology (director: G. Köhler). I expected that suggesting to install a departmental structure in the notoriously unstructured MPIIB would meet with a very positive response in the MPG's central administration and would thus increase the chances of the proposal. In addition, while it was not difficult to point out Köhler's main achievement, inventing the hybridoma technology, I felt it was necessary to stress that there were strong indications of continuous scientific productivity. I therefore mentioned in the letter Köhler's recent experiments on the selection of somatic mutants from hybridoma cells as far reaching and imaginative approaches to the future of mammalian cell genetics. Although Secher's and Milstein's papers on somatic mutants of the P3 cell line had appeared several years earlier, I did not realize at the time that Köhler's selection experiments yielded similar deletional and frameshift mutants without impact on the diversification process (see Chapter 7). Finally, I attached Nicolas Wade's article "Hybridomas, the making of a revolution" that had just appeared in the journal *Science*, in February 1982. In the article, Wade had argued that Köhler, although having had the primary role in the invention, was being neglected by the prize committees (see Chapter 6).

Of course, I had informed Köhler of my interest in bringing him to Freiburg and had tried to get an impression on whether he was willing to come. This is a game that is typical of academic appointments in Europe and some scientists are masters in playing it. Nobody in his good senses says no to an unofficial first inquiry relating to a possible offer of a professorship or institute directorship. Instead, one indicates interest with the aim to encourage the university or research organization to pursue the matter further and finally make an official offer. Once the offer is official, all advantages are on the side of the candidate and the real game can start. The details of the offer, including the personal salary, fringe benefits, scientific investments, number of personnel, and running budget are first maximized by negotiations with the future employer, always politely implying that one unfortunately has to turn it down if the demands are not met. Of course, there is a limit to what one can ask, and it is part of the art to correctly estimate this limit without overdrawing one's opportunities. Once the offer is optimized, one goes back to the present employer to let them know that one unfortunately has to accept the new opportunity unless the conditions of the present job are substantially improved, so they can compete with that of the new job. Many universities and research organizations are afraid of losing existing personnel, because hiring a new person always costs more than keeping the old, and therefore they will almost always make a counter offer. More often than not, there will be a second or even a third round of negotiations, as European universities rarely set a time limit for the candidate's decision. Some experts in this activity have managed to negotiate for up to 2 years before they decide. In the end, rather frequently candidates stay where they are, given the family situation, wife at work, children at school, etc. It is therefore of great importance to correctly assess the real intentions of a candidate before an official offer is made. Of course, one can never be sure until the contract is finally signed.

The first conversation in which I hoped to explore Köhler's interest in coming to Freiburg took place one afternoon in the spring of 1982. I had asked Köhler for an appointment by phone, indicating for the first time that he was being considered for a directorial appointment in Freiburg. A few days later we met in his laboratory at the Basel Institute and he took me to an open air cafe on the banks of the river Rhine in Basel which provided a relaxed and pleasant environment for this sensitive topic. The weather was nice and we could sit outside in the sun. We were having Swiss "Schümli" (espresso with foam) and Georges soon began teaching me a lesson in how to negotiate a scientific job offer. As expected, he indicated tentative interest in the opportunity but at the same time made clear that he was all but flabbergasted by the offer and that a lot of conditions would have to be met for him to leave the Basel Institute. The Basel Institute was beginning its second decade of existence and had joined the ivy league of the few worldwide leading centers in immunology. Melchers had succeeded Jerne as director and was successfully using his considerable talent in public animation to inform the scientific community as well as the general public about the breathtaking scientific achievements of the outstanding scientists working at the Basel Institute. Melchers let no opportunity slip by to praise the productivity of the anti-hierarchical organization introduced by Jerne in the Basel Institute and had succeeded to create an awareness within the institute of its unique excellence, not without reason. As a result, many of the scientists working there felt a strong self esteem and emanated a degree of superiority in their relations to others. Köhler was no exception and made no secret of the dim view he held of the Max-Planck-Institute in Freiburg, which he considered "provincial" and in need of development to become an internationally competitive research center. While we did not get into much detail on this occasion, I was left with the impression that it was worth pursuing the Köhler option further, but that it would take considerable persuasion to get him to Freiburg.

While this turned out to be true, I did not foresee the controversial discussion within the MPG that arose during the evaluation of Köhler's scientific and leadership qualification. The president of the MPG had accepted the nomination of Köhler and had passed the documents on to the chairman of the biomedical section assembly, Peterhans Hofschneider, to start the evaluation process. A committee was elected from the section members which contacted six leading immunologists, one German and five from abroad, for letters of evaluation. For obvious reasons, their names will not be disclosed. Only two of the reviews were entirely positive and recommended Köhler for the position without hesitation. The others contained partially very harsh criticisms. The mildest reservations concerned Köhler's leadership quality which most of the reviewers gave low marks. For example, one reviewer wrote "*I would not look to him as the type of person to inspire others, to create an independent group, to push German science into the international arena...*". More serious, however, was the general criticism of his achievements as a scientist. While most of the reviewers gave Köhler credit for his contribution to the monoclonal antibody work, several held the view that he had merely continued the work of Cotton, performing experiments that were "in the air" and would have been done by any other postdoc following Cotton in Milstein's laboratory. One reviewer wrote that he had "*not the slightest doubt that Köhler would not have done* (these experiments) *had he not been in Milstein's lab at the time*". The most serious criticism, however, concerned Köhler's scientific work before and after his time with Milstein. Here are a few quotes: "*Nothing that Köhler has done either before or after the monoclonal antibody work would itself justify his appointment...*", his "*previous work* (was) *respectable but uninspired*", his "*subsequent work essentially an extrapolation of the mAb procedures*", his "*work before* (was) *essentially routine ... neither more nor less than would be expected ... of a PhD student*", his work after "*very good but not great*". While most reviewers expressed their expectation that the monoclonal antibody work would eventially be honored by the Nobel Prize, there were serious doubts whether Köhler would share in it. One reviewer wrote: "*I am reminded of Nirenberg and Matthaei, although I hope and expect that the outlook for Köhler will be better than what appears to have transpired in Matthaei's case*". The reviewer referred to the Nobel Prize for the discovery of the genetic code that was shared by Nirenberg and not by Matthaei although everybody later agreed that Matthaei had significantly contributed. Matthaei, who had been postdoc with Nirenberg and later became Max Planck director in Göttingen, never recovered from his frustration and became a tragic figure in his later years. He was the only Max Planck director ever who had his directorial function taken away for lack of scientific productivity.

Most other evaluation committees would have refused to further consider a candidate who had received a majority of predominantly critical reviews. As directors of the proposing institute cannot be members of a committee evaluating their own candidate, I did not know about the poor reviews for Köhler when I was invited by the committee for questioning. I must have used the right arguments, for example when confronted with the concerns about Köhler's leadership qualities I explained that I was prepared to shoulder both the administrative burden as well as the future scientific development of the entire institute for several years to come and would keep Köhler's back free so he could fully concentrate on his scientific work. I also repeated my arguments defending Köhler's mutant selection experiments as innovative and ingenious approaches to mammalian cell genetics. Only much later I was able to read the reviews as well as the minutes of the committee which summarized the committee's conclusions on three main issues. First, the committee agreed that Köhler's contribution to the hybridoma technique, although it could not be assessed exactly, must have been significant. Second, in judging Köhler's subsequent scientific achievements the committee decided to follow the two positive reviews and my own judgement and concluded that they were "*original and interesting contributions*". Third, in reacting to the concerns about Köhler's leadership talents the committee discussed the possibility of an appointment as scientific member without directorial function. This type of appointment is possible in the MPG but exceptional and, if not purposely aimed for, potentially derogatory. In the end, it was not deemed to be a viable alternative and the committee unanimously resolved to suggest Köhler to the section assembly for appointment as scientific member of the MPG and director at the MPIIB. Rarely has a committee so obviously dismissed a majority of critical reviews upon evaluating a candidate for Max Planck directorship. The section assembly heard the case in its spring meeting in February 1883 and approved Köhler's appointment with great majority.

Following final approval by the senate the president informed Köhler of the offer on 14 March 1983 and the two agreed to meet for a first round of negotiations on 22 April 1983 in the president's office in Munich. The usual procedure is that the administrative staff of the president's office prepares a memorandum for the president's use during the negotiations, containing the details of the available resources including the number of free scientist slots, technical personnel, investments and running budget, laboratory space etc. The candidate has in most cases already been informed by the institute about the available resources and has been instructed about certain items that the institute may want to acquire in the context of the new appointment. The candidate therefore rarely agrees to all parts of the offer but is expected to request improvements and additions to the package, for example additional positions, an increase in the running budget, laboratory refurbishments or even a new laboratory building, etc. After listening to these requests, the president will instruct his aides to determine how far he can go to meet the requirements of the candidate, because any item that goes beyond what has been provided by the institute has to come from central resources of the MPG and will therefore reduce the flexibility of the president in negotiating other appointments. The first meeting may be followed by a second or third until a mutually acceptable agreement is reached. The negotiations also include the candidate's personal salary, which the president is to some extent free to increase above the basic tariff in order to compete with the candidate's previous salary or with counter offers the candidate may present from his previous employer, or with third party job offers the candidate may have at the same time.

In addition to president Lüst and Köhler, Dr. Edmund Marsch participated in the meeting, responsible officer for the MPIIB in the president's office. Dr. Marsch belonged to the old – now nearly extinct – type of scientific administrators who were genuinely interested in serving science to the best of their abilities. He had been in charge of the MPIIB for many years and knew all its details of the past and present, including Köhler's unfavorable reviews. Köhler, in contrast, did not know about the critical deliberations of his review committee and the more than lucky outcome in his favor. As a result, Köhler's considerable self esteem was unbroken and had perhaps even received a further boost by the Max Planck offer. While president Lüst and Dr. Marsch expected to meet a modest young man and the meeting to proceed in the usual way, i.e., to be concerned with positions and budget and the like, they were very surprised to hear that Köhler had so far not given the details of his appointment any thorough thought and did not wish to discuss them at this occasion. Instead, Köhler confronted them with a demand that he put forward as his main condition for accepting the Max Planck directorship: he wanted to be free to decide to retire at the age of 50, with the full pension that he would be entitled to at age 65, the legal retirement age. Moreover, Köhler informed president Lüst that he had presented his present employer, the Roche AG, with the same demand. Should Roche agree, he would turn down the Max Planck offer and remain in Basel. The minutes that Dr. Marsch prepared on this first meeting between president Lüst and Köhler contain no mention of Köhler's demand. They merely consist of the prepared document listing the available resources and state that Köhler had so far not commented on them. Obviously, Köhler's demand to decide about retirement at age 50 was considered so bizarre that Dr. Marsch decided not to lay it down in writing. Neither did he mention in the minutes that Lüst immediately pointed out to Köhler that arrangements of this type are out of the question in the MPG, which is part of the public service in Germany, no matter how badly the society might wish to hire a person.

When the meeting of 22 April 1983 was finished no date was fixed for a second meeting. The parties agreed that Köhler would let the president know when Roche had decided, one way or another, on his demand. The decision at Roche took until September 1983, when Köhler wrote a letter to president Lüst: "*Roche has offered Swiss Francs 20,000,–* (increase in annual income) *in response to my request to decide upon my retirement at age 50. I am now willing to consider to come to Freiburg. I hope that all other conditions can be settled quickly and generously*". Fritz Melchers recalls the way in which Roche decided on Köhler's demand as follows: Köhler had first informed Melchers, director of the BII, about his condition to stay with Roche. Melchers had inquired with Roche's financial department who came back to him with an astronomical sum that Roche would lose if it were to agree to Köhler's condition. Nevertheless, the matter seemed important enough to be dealt with at the highest level. Fritz Gerber, then CEO of the Roche AG, was informed upon the matter and invited Köhler and his wife together with Ursula and Fritz Melchers for cocktails in his house in Zürich, followed by dinner in a fine restaurant. Köhler repeated his demand, explaining at length how he envisaged the possibility that later in his life he would perhaps be dissatisfied with his work as a scientist and develop other interests. If that were the case, he would like to be in a position to decide what to do in his future without the pressure of having to earn an income. Gerber, who had met Köhler for the first time, listened to Köhler's deliberations without much comment, but Melchers recalls the irritated reaction of Gerber and the tense atmosphere for the rest of the evening. When everybody said goodbye Gerber took Melchers aside, saying "this guy must be out of his mind".

The negotiations that followed Köhler's agreement to seriously consider the Freiburg opportunity were all but easy. The substantial increase in Köhler's salary at Roche presented the Max Planck authorities with a serious problem. In order to compete, they had to go to the limit of the salary scale so that Köhler's income would be much beyond that of his peers in terms of age and experience, and would be in the range of that of the highest paid directors in the MPG. Moreover, Köhler demanded three scientist positions at the associate professor level, while the institute could provide only one of those. His argument was that he wanted his department to consist of four independent laboratory groups, one led by himself and the other three under the lead of advanced international scientists to be hired. With this structure he would guide the institute out of its present "provinciality". Altogether, he requested resources much beyond that which could be provided by the MPIIB and had to be carved out of central MPG funds. Nevertheless, president Lüst did not want this appointment to fail and Köhler got what he asked for. The decisions were taken quickly and Köhler signed his contract on 22 November 1983. He was employed at the MPIIB with a partial salary as of 1 January 1984, but the contract allowed him to retain his employment in Basel until the refurbishments of his new laboratories, covering the entire first floor of the MPIIB, would be finished, anticipated by about 1 March 1985.

# Chapter 11
# "Köhler's Max-Planck-Institute"

The Nobel Prize for Köhler, Milstein, and Jerne, announced on the first Monday of October 1984, not only changed Köhler's life but had a strong impact on the anticipated future development of the entire MPIIB. The Prize came as a surprise to many immunologists. There was first the fact that Köhler was included, which was not generally expected. Secondly, the fact that Jerne shared in the prize which was primarily given for monoclonal antibodies had not been expected either. In addition, many had foreseen that the hybridoma technology was eventially honored by the Nobel Prize, but few had expected it so soon. Research organizations go wild if one of their members is honored by the Nobel Prize. In the present case, three such organisations shared in this excitement as all three could claim parts of it. Monoclonal antibodies had been developed at the MRC`s Institute of Molecular Biology in Cambridge, so that the reputation of having hosted the honored research was theirs. Understandably, in England the excitement centered primarily on Milstein. The BII had full possession of Jerne, who had created a significant part of his theories, for which he was honored, in Basel. As for Köhler, both the BII and the MPIIB possessed a part of him, as he was on the payroll of both institutions, leading to a certain tension between the two institutions which included even the spouses of the two directors. During a reception in honor of Jerne and Köhler at the BII, which I attended together with my wife, Barbara, Ursula Melchers came close to physically attacking Barbara over their dispute to which of the two institutes Köhler's prize belonged.

Above all other things, the Nobel Prize is a feast for the media. This was particularly the case in Germany as Nobel Prizes for Germans have been rare in recent decades. As a result, Köhler was the main target in the coverage of the 1984 Nobel Prize for Physiology or Medicine in the German media, while Milstein and Jerne were only briefly mentioned. Not only was Köhler a German, he was also unusually young and had been made Max Planck director at this young age. A public debate arose whether Köhler had been appointed by the MPG before or as a result of having received the

Figure 19
Picture taken at a reception at the MPIIB in October 1984, in honor of Köhler's Nobel Prize. From left: MPIIB scientist Dietrich Hammer (back to camera), Otto Westphal, vice president of the MPG Benno Hess, Köhler, Otto Lüderitz, Klaus Eichmann.

Nobel Prize. From then on, the MPIIB in Freiburg was "Köhler's Max-Planck-Institute". The journalists had a great time also because the invention of monoclonal antibodies was much easier to convey to the public than, for example, the Mössbauer effect, and even better, was important in medicine. Many articles disseminated the unproven expectations on the use of monoclonal antibodies in therapies of cancer and other diseases. The term "magic bullet" was coined. In the public eye, Köhler became the new genius of biomedical science in Germany.

Köhler began working full time in Freiburg as of 1 March 1985. The first floor of the institute had been refurbished, a matter that had involved many intense and controversial discussions with the construction department of the MPG, who wanted to keep expenses low and therefore kept suggesting modifications to Köhler's plans. These discussions ended abruptly after the Nobel Prize, and the work was completed as requested by Köhler. The sudden end of resistance to Köhler's plans for laboratory refurbishment was only the first indication of a new submissive attitude of the MPG. From now on, anything that Köhler wanted was possible, and if Köhler wanted anything no other opinion counted. The authority of a Nobel Laureate is unsurpassed in Germany, apparently much in contrast to England. In the year after the Nobel Prize the Babraham Institute sent Milstein's wife, Celia, into early retirement at age 58 against her and Milstein's outspoken wish, in an act of "sexual discrimination" as Celia chose to call it. Nobody in Germany would have done anything against Köhler's will, not to speak of annoying his wife, in the years following the Nobel Prize.

When it came to assuming responsibilitiy for the entire institute and taking the lead in shaping its future, I soon began to realize how much I had misjudged Köhler's intentions. In the process of considering Köhler for directorshiop at the MPIIB, I had gained the impression that he was primarily interested in his own research and would be happy if others were to worry about the overall strategy for the future scientific development of the entire institute. By saying that I was prepared to continue in taking care of the latter tasks, I had calmed down the concerns of his review committee about the problems that the outside reviewers foresaw if Köhler were to be appointed to a position that required competent leadership. Either I had gravely misjudged Köhler's intentions or he had drastically changed after the Nobel Prize. The discussions on the Lüderitz succession, who was due to retire in 1988, began soon after Köhler's arrival in Freiburg and Köhler instantly took the initiative by suggesting that we try again to get Rajewsky. While Köhler knew about the long-term scientific concept submitted to, and approved by, the MPG in the context of suggesting Köhler's appointment, it now became apparent that his thinking went in entirely different directions. According to this concept, the Lüderitz department was to be taken over by a successor who was working on immunological defence mechanisms in infections. Infection immunology was, on the one hand, a modern field which was rather neglected in Germany but received increasing attention internationally. On the other hand, it seemed to be close enough to bacteriology which was still strongly pursued at the institute, and likely to be an area of integration among the old and the new. Suggesting Rajewsky meant to abandon this concept. However, Rajewsky kept being a strong candidate which was hard to argue against in scientific terms, particularly as he had been a prime choice before. Rajewsky somehow had been able to convince Köhler that he was now in a position that would make it possible for him to leave Cologne. For the MPG authorities it was hard to agree to the nomination, as it had never occurred that a directorship was offered to the same person again after it was turned down on the first try. However, Köhler was not to be disappointed and Rajewsky went again through the evaluation process, which he passed without any problems. What followed were again long months of negotiations and, as some of us had foreseen, a final turndown. Rajewsky was very experienced in turning down job offers after lengthy negotiations. One of his tricks was that when they were trying to set him a time frame for his decision, he would reply that with the proposed deadline he would regrettably have no other option than to say no. Moreover, he had developed a routine to make everybody believe how personally difficult it was for him to finally make the negative decision. Whether Köhler believed him or not, it was obvious that he was very disappointed with Rajewsky so that he later vetoed Rajewsky's candidacy to serve on the scientific advisory board of the MPIIB.

Needless to say, also the political caste got excited about Köhler's Nobel Prize. Already, in December 1984, the President of the Federal Rebublic of Germany, Richard von Weizsäcker, inquired with the MPG if Köhler would be willing to accept the "Bundesverdienstkreuz", similar to a "pour le mérite" order, of the state of Germany. Köhler declined to avoid what he called an "automatism" of honors coming to him. Instead, he offered to meet President Weizsäcker for a conversation, a suggestion that Weizsäcker chose not to respond to. The German minister for research and education, Heinz Riesenhuber, visited the MPIIB for an entire day, giving a demonstration of his erudicity but no extra financial support. Most consequential, however, was the enthusiasm of the Prime Minister of the State of Baden-Württemberg, Lothar Späth, about Köhler's prize. Späth, who had no academic education, had made his career to the leading elite in the Christian Democratic Party because he had excellent common sense, a well developed political instinct, and an inborn talent of leadership. Like many other politicians, he had subscribed to the widely held notion that the future of Germany depended on its intellectual and scientific achievements. Unlike most of his colleagues, however, he indeed made practical efforts to foster what he thought were outstanding scientific and technical developments. In doing so, he tended to focus on matters that were easy to sell to the public, such as prominent individuals or single spectacular developments, whereas broad investments into universities or long-term educational or research programs were of less interest to him. Already in the summer of 1985 he let Köhler know that he was willing to support his work by a substantial investment and invited him for a visit in his residence in Stuttgart, the Villa Reizenstein. Köhler asked me to come along as acting director of the MPIIB. The meeting was held in the evening over dinner, in the presence of several of Späth's aides. The conversation was essentially a monologue of Späth, who talked at length about his visit to Silicon Valley and his successful efforts to acquire a Cray III supercomputer, for use by the various universities, research organizations, and industries in Baden-Württemberg. The substantial costs of 75 million Deutschmarks had been carved out of the state budget, but he was convinced that a multiple of the sum would come back to the state in terms of user fees. Several years later I remember receiving letters which were directed to potential users trying to sell Cray III computer time, and there were rumors that only a small number of users had ever registered. Only at the very end of the evening Späth came to speak about the purpose of the meeting, his wish to support Köhler's research. Köhler briefly presented our plan to build a new laboratory building and to establish a new department as well as several independent junior research groups at the institute, which would require a temporary increase of the institute's personnel by about 25 scientist and technical positions. Späth reacted very positively and adjourned the meeting by saying that he would look forward to receiving detailed plans in writing.

In addition to the traditional department under the lead of a director, the Max Planck Society supports independent junior scientists, providing them with three to five positions for scientific and technical collaborators and sufficient funds to carry out research programs of their own choice and not necessarily related to the main subject of the institute. The positions are advertised internationally and the typical successful applicant has spent a productive postdoctoral time in a prominent laboratory in the US, his work being documented in several publications in high impact journals. The resources and independence associated with a junior research group provide opportunities that are hard to match by any other structure and the number of applicants frequently exceeded 100 per advertised position. The only disadvantage was that contracts were limited to 5 or 6 years with no possibility of extension. The junior group structure was meant as a breeding ground for the recruitment of scientists at the full professor or directorial level in Germany. While it indeed served that purpose initially, the system deteriorated with time and lost its reputation as a promising career start, mostly because there were not enough follow-up positions available in Germany. In 1986, however, the career expectations of junior group leaders were still good, and many Max-Planck-Institutes made efforts to establish junior groups.

The plan to suggest to Späth the financing of a new laboratory building with corresponding personnel had been developed between Köhler and myself by mutual agreement, initially in the context of the attempts to get Rajewsky to Freiburg. Any substantial new research activities in the MPIIB required both additional space and positions. The collaborators of Westphal and Lüderitz who were still at the institute had permanent contracts so that there were no resources worth speaking of, neither laboratory space nor free positions, available in the institute. Making resources available would have meant withdrawing them from existing groups, a strategy that can be quite destructive to the overall atmosphere in an institute. The Späth investment came thus at the perfect moment, solving this problem and, in addition, providing the resources of a new department as well as junior groups at the MPIIB. Most of the temporary positions in the package would be returned to the government with the retirement of the present occupants, so that in the end it was mainly the new building that was to be financed. The plan was enthusiastically supported by the new president of the MPG, Heinz Staab, then negotiated between the Max-Planck Central Administration and prime minister Späth's financial department, and finally approved by all parties in 1986.

When it came to the scientific substance for the new capacity after Rajewsky's turndown, I tried to revive the original plan to establish infection immunology at the MPIIB. My favorite candidate for this activity was

Figures 20, 21, 22
Köhler giving tours of the MPIIB to politicians: Federal minister of science and technology Heinz Riesenhuber, Baden-Württemberg prime minister Lothar Späth, and Späth with Freiburg's mayor Rolf Böhme.

Stefan Kaufmann, who had been a postdoctoral fellow in my department in the MPIIB for the first 5 years of its existence and had then accepted an associate professorship at the university of Ulm. Kaufmann was interested in intracellular bacteria, using infections of mice with *Listeria monocytogenes* and *Mycobacterium tuberculosis* as models. He had already made important contributions to understanding the immunity to these infections, including the discovery of a novel role for cytotoxic T cells in resistance to intracellular microorganisms. Realizing that Köhler was ignorant about the field of infection immunology and would therefore likely not appreciate Kaufmann's work, I had written to a selection of leading international experts asking for suggestions who they would choose as their prime candidate for a department of infection immunology at the MPIIB. Not having suggested any names in my letter, Kaufmann came out as the clear winner of this international search. Hoping that this would convince my colleagues I presented the results at a meeting of our internal search committee. Köhler congratulated me to the professionality of the search but was not impressed by the result. While this was halfway expected, to my great disappointment the bacteriologist members of the committee could not be convinced either of Kaufmann as a candidate. As a result, infection immunology was no longer pursued as an activity for the future of the MPIIB. Soon thereafter, Kaufmann was promoted to full professor in Ulm, and in 1993 was appointed founding director of the new Max-Planck-Institute of Infection Biology in Berlin. He is now a leading international authority in his field. The outcome of the Kaufmann debate was a further milestone in my realizing that trying to get the better of a Nobel Laureate is a waste of time.

Having abandoned infection immunology as a future activity of the MPIIB left us with the task to define other subjects to fill the new building. In this type of situation, when significant new resources and potentially a new scientific orientation are at stake at an institute, the Max-Planck Society appoints a "stem-committee" to deal with the matter. Stem-committees are appointed by the section assembly in much the same way as committees in charge of directorial appointments, but their mandate includes the long-term scientific strategy of the institute in addition to deciding upon the candidates in question. As a result, a stem-committee may itself suggest directorial candidates who in the normal procedure are nominated by the institute. Like any other committee, a stem-committee is elected by a voting process in a section assembly meeting. The section members who serve as candidates in the election are in part suggested by the institute in question prior to the meeting, but can also be spontaneously nominated from the floor during the meeting. Erudite section members who often speak up and are well known to the assembly are frequently nominated and elected, independent of their field of expertise. As a result, there is a certain unpredictive randomness in the composition of section committees. At the time when the stem-committee for the MPIIB was elected, it so happened that the field of developmental biology was very popular in the MPG. A number of young developmental biologists had been recently appointed as directors, among others Christiane Nüsslein who later received the Nobel Prize for her pathbreaking work on the development of *Drosophila*. She and several others had been appointed as directors to the Max-Planck-Institute of Virology in Tübingen, which had thus changed its entire scientific orientation and a few years later was renamed Max-Planck-Institute of Developmental Biology. Nüsslein and her colleagues represented the young progressive fraction in the section assembly that were nominated for almost any committee election. An important fact also was that there were very few immunologists in the section at the time. Köhler and myself, as members of the institute at stake, could not serve in the committee. Norbert Hilschmann was never nominated because he always argued against everybody else, and Jan Klein did not attend section meetings and was known to refuse committee memberships. The result was that the stem-committee of the MPIIB came out without any immunologist and with a strong bias for developmental biology.

When the stem-committee met in Freiburg for the first time in 1987, they asked Köhler for his ideas for the future of the MPIIB. At the time Köhler was toying with an interest in hemopoietic development, improvising in front of the committee that in his view the common goal in the new building should be to establish long-term tissue cultures of hemopoietic stem cells so that their properties could be studied. While this was mostly a transitory interest that Köhler never more than marginally pursued in his own work, the committee members took it to be the result of the deep and thorough thought process of a Nobel Laureate. Köhler's remarks were favorably received by the developmental biologists in the committee, as hemopoiesis is a subspecialty of developmental biology. The subject also was sufficiently related to immunology, as the cells that mediate immune functions – the lymphocytes – arise by differentiation from hemopoietic stem cells, as do all other cells circulating in the blood. Indeed, the hemopoietic stem cell is of primary interest also to clinical immunologists. It is the cell that gives rise to the reconstitution of all blood cells following bone marrow transplantation, a procedure used in the treatment of leukemias and recently also of immunodeficiencies and autoimmune diseases. At the time it was known that hemopoietic stem cells existed in the vast mixture of cells comprised in the bone marrow, but their properties were largely unknown and nobody had a clue which of the cells in the bone marrow was a stem cell. Köhler's suggestion, spontaneous as it was, was therefore persuasive and an attractive subject for a Max-Planck-Institute. The committee was impressed by Köhler's suggestion but found that defining a field as narrow as hemopoiesis would restrict the search for candidates more than would be desirable. Therefore, the committee resolved that the extension of the MPIIB in the new laborytory building would include research activities in the field of developmental biology.

In the same year, it became known by the grapevine that Davor Solter was seeking a position in Europe. Of Yugoslavian decent, Solter had studied Medicine in Sweden and had spent his entire scientific career in the US, in recent years in the Wistar Institute, Philadelphia. There he directed a big laboratory together with Barbara Knowles, his partner both professionally and in private life. More radically than in other US research outfits, in the Wistar Instute scientists had to finance their entire expenditure out of their personal grants, including everything from their own salaries to the toilet paper. As a result, as their group grew, Solter and Knowles were forced to spend more and more time on grant writing, an automatism that to Solter seemed like a trap in which he would be caught for the rest of his professional life. Solter therefore privately indicated to friends and colleagues that he would consider offers from elsewhere, including from Europe where scientists depend to a lesser degree on the acquisition of grants. As soon as the word got around Solter, who enjoyed a superior scientific reputation, received multiple inquiries from institutions in the US and also from Europe. Solter was one of the first to try cloning of mammals by nuclear transfer experiments, at the same time as the Swiss cell biologist Karl Illmensee who had raised considerable public interest when he reported the first successful embryonic cloning experiment: he had removed the nucleus of a zygote, the earliest embryonic stage of the mouse, injected the nucleus of a cell derived from a more advanced embryonic stage, and derived living progeny after implanting the reconstructed zygote into the uterus of a pseudopregnant female mouse. Removing and replacing the nuclei were done by two consecutive injections using a microinjection device, one for sucking up the nucleus to be removed and the second for injecting the replacing nucleus, thus penetrating the cell membrane and surrounding layers of the zygote twice. In trying to repeat Illmensee's published method of nuclear transfer Solter completely failed. Illmensee refused to demonstrate his technique during a visit to the Wistar Institute, claiming that Solter's micromanipulation equipment was in bad condition. Not much later, after giving up his attempts to reproduce Illmensee's technique, Solter was successful in creating cloned embryos using an approach that removed the nucleus of the zygote by sucking it up with a piece of membrane, followed by cell fusion for getting the foreign nucleus into the zygote. As a result, scientists working on embryonic cloning began to lose trust in Illmensee's published experiments. Using nuclear transfer experiments Solter afterwards discovered the phenomenon of "genetic imprinting", again by disproving experiments previously published by Illmensee. While it was clear that the sex-chromosome Y was only present in the male, it was previously thought that the genes on all other chromosomes, the autosomal genes, were the same in male and female. In support of this concept, Illmensee had published papers in which he claimed that he had obtained living progeny from enucleated zygotes reconstituted either with nuclei containing two sets of female chromosomes, so-called partenogenetic nuclei, thus generating entirely gynogenetic mice. These experiments seemed to prove that normal offspring inherited two identical sets of autosomal genes from father and mother. Again, Solter repeated these experiments and found that zygotes with either two sets of female or two sets of male chromosomes never developed further, whereas control zygotes having received nuclei containing one set of female and one set of male chromosomes gave rise to living offspring. Thus, Solter showed that male and female genes were not identical, some autosomal genes could only be provided by the father and others had to come from the mother. He thus had discovered the reason why in normal life the generation of viable offspring requires a father and a mother and cannot be achieved by combining the genomes of two males or two females, which should be possible if the male and female genomes were identical. He concluded that male and female organisms altered the structure of their autosomal genes in different ways, they imprinted their genes. While Illmensee became a tragic figure who lost his position and scientific credibility, Solter became a worldwide authority in reproductive cloning. Solter today: "Sometimes I think that Carl did some microinjection experiments, mixed up his embryos, and thought that it worked. He then published what in his mind worked once, claiming that it was a reproducible result that he had obtained many times." Imprinting was later shown to be due to inactivation of certain genes by epigenetic modification, a process that involves DNA-methylation and affects, in addition to the X chromosome, different sets of autosomal genes in male and female germ cells. As a result, certain autosomal genes in the male or female genomes are inactivated and these genes have to come from the opposite parent.

Solter was alerted to the opportunity in Freiburg by Rolf Kemler, then leader of an independent research group in the MPIIB. Kemler (see Chapter 14) knew Solter as a colleague for a long time, and had heard of Solter's interest in finding a new position, possibly in Europe. During a casual conversation sitting on the lawn in Cold Spring Harbor they agreed that Kemler would pass Solter's curriculum vitae and list of publications on to Köhler with a view to get him interested in Solter as a candidate for director at the MPIIB. Köhler seemed to be open to the idea and all of a sudden Solter was circulated in the MPIIB as a directorial candidate. He was a first rate scientist and it was clear that, following the Harnack principle, there could not be any doubt that Solter would pass the MPG evaluation process without problems. He was a developmental biologist and the stem-committee had selected this field for the MPIIB, not to speak of the overall excitement that Solter's appointment would generate among the young developmental biology fraction in the biomedical section assembly. The fact that Solter would mean a significant deviation from the subject of hemopoiesis, was of minor importance if one considered the unique opportunities associated with this prominent appointment. Solter visited the MPIIB and developed a serious interest in joining it, as grant writing was no issue and the structure of the departments in the institute agreed with Solter's dream to work with a small number of peers at equal scientific level rather than having to supervise a large number of junior scientists. In preliminary discussions it was agreed that the institute would provide seven scientist positions to Solter, equal to that of my department and one less than that of Köhler's.

Köhler had in the meantime become acting director of the MPIIB and was in the process of organizing a symposium in which the shortlisted applicants for the junior group leader positions were to present their work to the members of the institute and to the stem-committee for final selection of three group leaders. There were quite a number of excellent presentations, and Köhler spontaneously developed the plan to establish four junior groups rather than three. Köhler then tried to carve the two positions required for the fourth junior group out of Solter's package, but Solter firmly insisted on the original agreement. Finally, Köhler gave in and suggested Solter to president Staab, who approved of the suggestion and passed it on to the section chairman, Klaus Kühn, for evaluation. The stem-committee of the MPIIB was much in favor of Solter who predictably received only highly positive international reviews, and the section assembly approved Solter's appointment with a large majority. Solter accepted the offer after only a brief period of negotiations, although the MPG could not come up with an arrangement that would persuade Barbara Knowles to come along. The two had anticipated a position for Knowles at a level equal to that of a director, but the MPG did not offer more than an associate professor. Knowles decided to stay in the US and became director of the genetics program at the Jackson Laboratory. The two remained a couple, however, leading a successful long-distance relationship.

Köhler spent much of his time planning the new laboratory building. A contest had been advertised, a number of prominent architects had submitted their plans, and a local architect came out as the winner. The building was a large cube of three stories of laboratories that surrounded a wide inner court which opened up over the height of the upper two floors and received daylight through a large glass roof on top. The design excited everybody including architects and scientists, the latter because the floor of the inner court, the "atrium", seemed to provide the ideal structure to facilitate interactions among the personnel working in the laboratories surrounding it. The building later received a prominent German architectural award, the "Hugo". In the weeks and months of detailed planning that followed Köhler participated in almost any meeting, giving attention to any matter including the smallest of details. I remember long discussions about the directions in which certain doors would open and on which side of the door the handles would be attached. A conflict arose between Köhler and the architect about the height of the shelves on top of the laboratory benches, which Köhler wanted to extend up to the ceiling in order to generate storage space whereas the architect wanted them low because he had

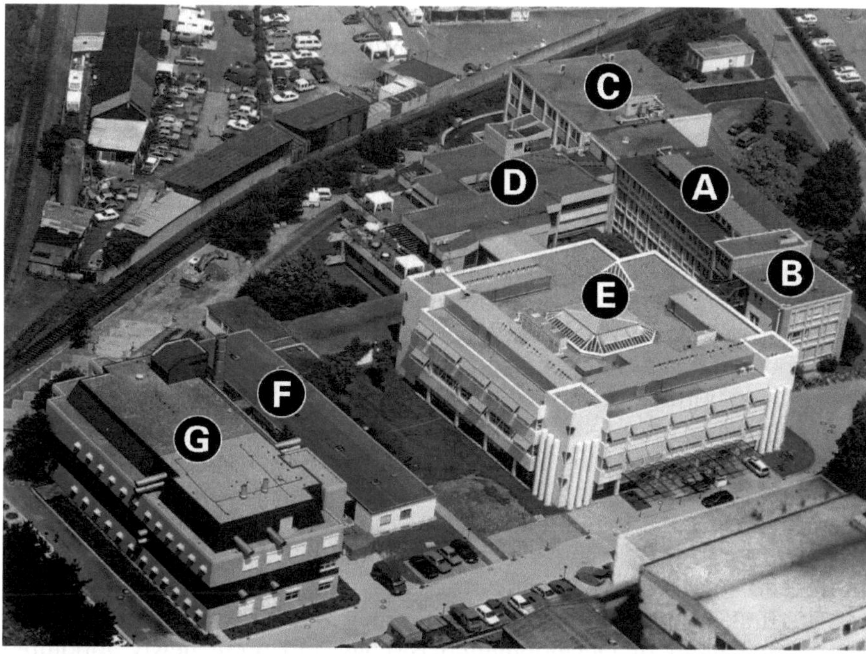


Figure 23
Aerial photograph of the Max-Planck-Institute of Immunobiology in 1995. The various buildings of the institute complex were constructed sequentially over several decades starting with A, the first laboratory building erected as the Dr. Wander Research Institute in 1957. Part of its conventional animal quarters (F) are still in use today. Buildings B and C were added on both sides in the late 1960s to increase laboratory space. In the mid-1970s an infrastructural tract (D) was added to house the library, lecture hall, cafeteria, and administration offices. The specific-pathogen-free animal house (G) was built in 1980 and later extended to its present size. The square shaped structure in the center (E) is the laboratory building donated by the state of Baden-Württemberg in recognition of Georges Köhler's Nobel Prize. It was finished in 1989, and was used to accomodate the developmental biology activities and the Spemann Laboratories.

planned glass partitions between laboratories in order to create the impression of open space. In spite of many such small controversies, Köhler was unusually amiable in his dealings with the architects and one had the impression that bringing up this building made him really and deeply happy. Nevertheless, and to the deep disappointment of all who had contributed to the financing, planning, and construction of the building, Köhler never seriously considered moving his own laboratories into it. When he was asked for the reasons, he would say: "I want to use the building to attract excellent new scientists to join the MPIIB, the best possible".

Construction of the new laboratory building was finished in 1989 and the inauguration was celebrated with a scientific symposium and a ceremony, attended by prime minister Späth, MPG president Staab, and a number of representatives of the city of Freiburg, the University, local industries, etc. Everyone gave speeches and congratulated one another on their mutual achievements. The event resulted, however, in the beginning of a longlasting fallout between Köhler and the mayor of Freiburg, Rolf Böhme. Böhme had come with a painting that he had planned to give to Köhler at the end of his speech as a present by the city of Freiburg, expecting that Köhler would unwrap it and appreciate it in front of the audience. Köhler disappointed Böhme first by having reserved a seat for him in the second row, which Böhme ignored placing himself next to Späth in the front. In addition, there was no speaking time allotted to Böhme and Köhler treated the painting by placing it, still wrapped, behind himself against the wall without giving it any attention. Köhler certainly had not annoyed Böhme on purpose, it just did not occur to him that Böhme could find it important to be seated in the front row or have his present unwrapped and appreciated in public. For years to come, Böhme would never again meet with Köhler even though he might have wanted to do so for reasons of public attention. This only ended when Roche indicated their tentative interest to build a new research building in Freiburg, a plan in which Köhler was involved as an initiator and advisor, and which was strongly endorsed by mayor Böhme. In the end, there were too many obstacles and Roche lost interest.

The first groups that moved into the new building was Davor Solter's new Department of Developmental Biology and three newly appointed young promising scientists leading independent junior research groups. To document their independence, junior group laboratories often receive names that differ from that of the host institute, in this case "Hans Spemann Laboratory", after the Nobel Prize winner and developmental biologist who worked in Freiburg in the 1930s. All of a sudden, the MPIIB had nearly doubled in laboratory space, personnel, and scientific scope.

## Chapter 12
# Human relations

For the first few years of his appointment in Freiburg Köhler and his family continued to live in Efringen-Kirchen, a small village about 60 km south, nearer to Basel than to Freiburg. They had bought a house there after Köhler had returned from Cambridge to become a permanent member of the BII. The house was a nondescript farmer's house, not very old but not new either, next to a small street in the center of the village. They had it refurbished so that one could see many of the old wooden beams that supported the structure of the walls and ceilings, the space in between painted a sheer white. There was rather sparce furniture, mostly unpainted wood of light brown color, and no carpets covering the wooden planks of the floor. Apparently the house had insufficient rooms for all family members to sleep in separate rooms, and the Köhler family slept for some time all in one room. When this turned out to be untenable for both the adults and the children, Köhler consulted an architect to solve the problem. The architect, Mr. Horst Linde, was a celebrity in his field in Germany, and not used to be concerned with trivia of this sort. The matter was discussed at some private invitation when both the Köhlers and the Lindes were present. Mrs. Linde recalls the story as if Köhler had asked her husband for help with nuptial problems arising if everyone sleeps in the same bedroom. Linde advised to install some additional walls but did not want to engage in designing detailed plans for Köhler's house. Rather than adding walls to his house, Köhler bought an apartment in Freiburg and moved there in the early 1990s.

While in Efringen-Kirchen, Köhler kept friendly but distant relationships to the farmers living in the village. Visitors who did not know his address just had to ask any person on the street for Köhler's house and were usually given correct directions. He was known to the villagers from some of his habits, for example he liked to roller skate with his children on the shallow slopes of the streets in the village. He also bought a small tractor, the type that owners of vineyards use to drive between the rows of grapes, and enjoyed driving it on the country roads around the village.

Köhler was a contemplative type. When asked what he liked to do in his free time, he would mention driving the tractor, or sitting in his garden to observe the flowers and vegetables planted by Claudia, who was a keen gardener and was proud of raising healthy food for her family, without artificial fertilizers or insecticides. Köhler also liked to play pingpong. I once did a match with him and was given a demonstration of his considerable skills. He mostly hit slow balls, but he could give the ball such a wicked spin that it was hopeless for his opponent to try to keep it in the game.

After moving to Freiburg, Köhler soon bought an old mill in the Alsace where he could continue his contemplative leisure activities on weekends. The mill had been unused and uninhabited for a long time and needed a lot of handiwork to be made inhabitable, which Köhler hoped to do largely by himself. His apartment in Freiburg was in a busy city street very near the city center, and had no garden or terrace. It was on one of the upper floors of a five-story apartment house built at the end of the 19th century and attached on both sides to the next house. Köhler himself was not all that happy with his buy, he complained that the place was noisy and dark. Claudia, however, seemed to like it and still lives there until today. In the absence of a garden, Köhler selected one of the city squares for his outdoor life in Freiburg. On Sunday afternoons, one could meet Georges and Claudia sitting on a small stone wall on Augustinerplatz, a cobblestone square not far from the Köhler's appartment in the city's pedestrian zone which is popular with students and families to enjoy a leisurely chat in the sun.

Köhler and his wife were very much attached to one another and to their children, giving the impression of a closely knit family. Institute parties and celebrations Köhler nearly always attended with his wife and often with some or all of his children. He would sit with his wife all evening, rarely leaving her alone to communicate with colleagues or collaborators. Social communication of the couple was to a large extent a task performed by Claudia, while Georges contributed short sentences and remarks. In his relationships to women Köhler was rarely if at all flirtatious or charming. Rather he would interact with women in the same way as with men, friendly but to some extent withdrawn and keeping a distance. For all one knows, Köhler was a faithful husband and did not have any affairs, even at scientific conferences that he only occasionally attended without the company of his wife. This definitely required some degree of self-control and resistance, as he was an attractive male and a celebrity, and certainly had no shortage in women trying to get in contact.

In contrast to women, Köhler was in no way indifferent to money, he rather was of the opinion that he and his family deserved to live unbothered by financial limitations. Although he had successfully negotiated a top level salary with the MPG, and received considerable consultancy fees from Roche in the context of a cooperation agreement, he still tried to make more, and used whatever opportunity came along. Shortly after Köhler's

Figures 24, 25, 26
Köhler attending social events in the MPIIB: Trying on a Japanese hat; mixing with children receiving gifts from Father Christmas; in Bavarian outfit getting ready for a milking contest.

arrival in Freiburg the publishing company Spektrum der Wissenschaft, publishing a German version of *Scientific American* in addition to other scientific journals and books, had asked Köhler to edit an issue on immunology in their series *Science Reader*. This series consists of collections of articles in a certain field that had appeared over the years in different issues of *Scientific American* and are now assembled in a single issue of *Science Reader*. A prominent scientist is invited to act as editor and to write an introduction. Köhler had asked me to be his co-editor and together we composed a fine issue of *Science Reader* that sold rather well and was popular with students who used it as an advanced textbook in immunology. Probably because of the success of the first immunology reader, Spektrum asked Köhler again nearly 10 years later to edit another issue on immunology. Köhler, who apparently had been disappointed with the meager royalties of the first edition, replied that he would do it for an advance honorarium of 10.000,– Deutschmark. This did not agree with the policy of Spektrum and so there was no second immunology reader.

Köhler would arrive at work not before 10 am. He had bought a bright red Peugot 306, a very fast little vehicle, to minimize the driving time between Efringen-Kirchen and Freiburg, but also after moving to Freiburg he would not come in earlier. He had installed his office in Westphal's former secretariate, a rather small room of about $4 \times 4$ meters, whereas Westphal's large office had become the meeting and seminar room of his department. Strangely, Köhlers secretary sat in an even smaller room behind his office, so that visitors would first enter his office before seeing his secretary. The doors of his office to the hallway and to the meeting room were always open and his collaborators had free access to him at all times. The meeting room was always equipped with coffee, soft drinks, and snacks, and people were sitting there at any one time chatting. It was hard to imagine how Köhler succeeded in getting his work done under these conditions, but he somehow managed. Köhler had entirely given up working at the bench in Freiburg, although he had still done so full-time in Basel. Nevertheless he studied the original experimental results of his collaborators in depth, brooding over them for hours on end by himself or discussing them with the person who had generated the data. Most of his co-workers, particularly the young students and postdocs in his group, were very fond of Köhler. He had a productive way of interacting with them and liked to listen to unconventional scientific ideas, the wilder the better. He never pulled ranks on his people, but easily managed to exert authority by his experience and self-assertive personality.

While his co-workers mostly had the benefit of Köhler's full attention, this was not always the case for other members of the MPIIB or outside visitors. If one entered his office for discussing some issue he would sit at his desk with his back to the entrance, mostly in the process of looking at some experimental results, an autoradiograph of an electrophoresis or a table of numbers of some sort. He would ask his visitor to sit down but to give him a few minutes until he was finished with his thoughts. The few minutes could well extend to a quarter of an hour or longer, during which time he would continue silently contemplating his documents while the visitor would just have to patiently sit there and wait. When he finally gave his attention to the visitor it was some sort of a lottery if a conversation was going to arise or if the encounter would continue mostly in silence. For example, Köhler was once visited by the science writer, Ernst Bäumler, best known for his authoritative biography of Paul Ehrlich. Bäumler was considering to write about Köhler and, as Köhler knew neither Bäumler nor his book about Ehrlich, I had helped Bäumler to arrange an interview with Köhler. Köhler agreed to see him but the meeting turned out to be highly frustrating for Bäumler. Later Bäumler told me that Köhler had just sat behind his desk and had responded to most of his questions simply by saying nothing. Needless to say, Bäumler abandoned his plans to write Köhler's biography.

Köhler was very good at frustrating others. One example of my own experience was the affair about "art-at-the-building". In the years when public finances were still flowing freely in Germany, the German government had issued a recommendation that 1% of the cost for major public buildings should be spent on art. For each prominent public building, an artist was to be selected and commissioned to prepare a work of art for permanent display in association with the building. Selection of artists and their works was left to the discretion of the users and architects of the building, and there was no end to the variety of possibilities that could be entertained. During the planning of the new laboratory building we had internally discussed the matter but nobody had so far come up with a good idea. On a visit to Israel sometime later I had the privilege to meet the internationally famous Israeli artist, Dani Karavan, who specializes in artistic landscaping. One of his most famous works is an environment termed "Passages" in Portbou in the Spanish Pyrenees, in honor of the philosopher, Walter Benjamin, who died there in 1940 trying to escape from the Nazis. Earlier he had done another spectacular job in designing a flight of terraces descending from the Roman–Germanian Museum in Cologne towards the river Rhine, termed "Way of Human Rights". I mentioned our pending art-at-the-building option in Freiburg and Karavan became interested. He knew Freiburg and loved the little brooks called "Bächle" that run through the old city center in Freiburg, one of the historical hallmarks of the town. We agreed that Karavan would visit the MPIIB a few months later when he was due to travel to Geneva for another commission. I informed Köhler about the invitation but could not get him excited, as he did not know Karavan or any of his works. Nevertheless he agreed to Karavan's visit and also to meet him. Karavan inspected the construction site and the plans of the building and to my great excitement promised to consider the matter and send us suggestions for the design of the access area in front of the new building. One of his spontaneous ideas was to base the design on the

"Bächle" motif to emphasise the relationships of the institute to Freiburg. Afterwards a group including Karavan, Köhler, the architect, a representative of the MPG construction department, and myself, went to a restaurant for lunch. In the meantime, Köhler had obviously decided that he disliked Karavan and, to my great embarrassment, ignored him in his most unpolite manner during the entire lunch. Karavan kept his posture but did not pursue the matter further. Indeed, nobody took any further initiative in terms of art-at-the-building, and so the building remained artless until today.

As a Nobel Laureate, Köhler received hundreds of requests from individuals or organizations for support of various charities and initiatives. He readily signed many such public appeals, for example to ban chemical weapons, to support activities against tropical diseases, various peace activities, etc. On the other hand, he was a lot more selective when it came to actually supporting initiatives by financial contributions or actual work. His papers contain only two monetary contributions, one to a women's house in Freiburg providing shelter to battered women and their children (DM 900,–) and the other to an initiative against biological warfare (USD 50,–). When he was asked to write a short article in support of suffering children in the third world, he did not reply. An exception was a letter that he wrote to the presidents of various national academies of science to solicit their support against the eradication of villages under the Ceaucescu regime in Romania. There were only a few answers. Köhler resisted numerous attempts by Israel Halpern, chairman of the Canadian Campain of Scientists and Scholars, to be enlisted in various human rights campains, with one exception, a campaign for Nelson Mandela. Köhler was also rather selective in joining boards of trustees or advisors, for which he received many invitations. One organisation that succeeded in enlisting him was the Socienty for AIDS Research, founded by the retrovirologist Georg Hunsmann, director of the German primate center in Göttingen. Hunsmann had also persuaded Nobel Laureate Manfred Eigen, so that Köhler had proper company on their board. The society administered much of the AIDS research support of the German government, but was rather heterogeneous and included in addition to basic scientists also physicians worried about how to advise their patients in safer sex. Köhler received multiple letters from a Dr. Hartmann, general practitioner in Munich, who wanted to enter into a debate with Köhler on how to reduce the infection risk for pregnant women. Köhler, who certainly could not care less than for this subject, either did not answer the letters or did so very briefly. The way Köhler answered most letters, not only Dr. Hartmann's, was by writing a few sentences by hand on the free space of the original letter he had received, and passed this on to his secretary for typing. This technique of answering letters meant that the length of his answer was limited by the space left on the letter he received, i.e., the longer the letter the shorter was Köhler's answer. After a few years, Köhler resigned from the board of the AIDS society, officially in protest of a reduction in government support for

AIDS research in Germany. More likely, he was happy the German government gave him a reason for his resignation. The Society for AIDS Research, if it still exists, never assumed an important role in the German scientific scene after the cutback in public support.

Other than in most of these activities, Köhler showed considerable motivation in his engagement for the German Museum. The German Museum, housed in a large 19th century building near the river Isar in Munich, contains an impressive collection of technical inventions covering the history of technology from prehistoric times to the atomic bomb and space craft. In the 1980s the German Museum was planning to established a new branch in Bonn, focussing on the scientific aspects of technology, including presentations of German Nobel Prize winners with their scientific achievements. The designated director of the new branch, Dr. Peter Frieß, visited Köhler in Freiburg and asked for help with the display describing the invention of the hybridoma technology. Köhler gave his laboratory notebook containing the Cambridge notes to the museum for permanent display. He also joined the scientific advisory board of the museum, attending its meetings regularly, and met with members of the staff making suggestions on how to design the displays. Members of the staff recall that he was sparkling with unusual ideas which he threw into the discussions spontaneously and in a persuasive fashion, a bird of paradise among the other rather formally minded advisory board members. There was a discussion going on in the board on how to best display scientific inventions and one of the concepts under consideration was to create "rooms of ideas", in which the discoveries leading to Nobel Prizes should be presented. In a draft text describing the concept somebody had written that the displays were to emphasize "*the spontaneity and playfulness of the research process ... as a means to overcome the chaotic aspects of human nature*". Köhler studied the text in detail, making many handwritten notes in it including an "*uff*" next to the above sentence. Köhler suggested to install telephones with television screens in each of the rooms of ideas which visitors could use to call Nobel laureates to ask them scientific questions. Indeed, several Nobel laureates participated in this activity, so that the idea was temporarily put into practice. Together with Frieß, Köhler made plans to produce a movie about monoclonal antibodies. They worked hard to raise funds for the project approaching, among others, the Bavarian minister for science and education, Dr. Hans Zehetmaier , the president of the MPG, Prof. Zacher, and Fritz Gerber, CEO of Roche AG. In their responding letters they all praised the project, wishing the best of luck, but apologized politely that financial support could not be provided. As a result, the movie project had to be abandoned. Köhler also suggested to design an educational board game in which monoclonal antibodies battle with viruses. It was completed after his death and is now on display in the museum for active use by visitors.

Köhler got surprisingly few requests to evaluate other scientists. Of the few reviews he wrote, most were critical to negative. For example, he was asked to write a letter in support of the nomination of Antonio Coutinho for the Premio Pessoa Prize, a highly prestigious science award in Portugal. Coutinho, who had recently been appointed director of the Gulbenkian Institute in Porto, had worked in the BII for several years and partially overlapping with Köhler. Of Portugese origin, he had started out as a doctoral student with Göran Möller in Stockholm before experiencing a rather spectacular career, occupying leading scientific positions in several European countries. He had cooperated with many colleagues, and was at times one of the most prominent and influential European immunologists. Coutinho had worked on different subjects in immunology with varying success, including the idiotypic network and its role in immune regulation and autoimmune diseases. Indeed, he had continued working in this field after it had been deserted by most other immunologists because its reputation as a serious scientific subject had faded away. Köhler, never a believer in the network theory, refused to write a letter of support for Coutinho. Another colleague who did not enjoy Köhler's support was Jörg Reimann, who was up for promotion to professorship at the University of Ulm. Reimann had done rather solid but somewhat unspectacular work on antiviral vaccines, having found novel ways to include viral proteins into vaccines such that cytotoxic T cells would be induced which are important in antiviral immunity. His work did not impress Köhler, who was not interested and largely ignorant in infection immunology (see Chapter 11), and he did not see fit to recommend Reimann for promotion. A grant application of Hans Georg Zachau to the Deutsche Forschungsgemeinschaft received only a halfhearted recommendation. Zachau, chairman of biochemistry at the University of Munich and a scientist of some prominence in the German academic scene, had for many years worked on the mouse and human immunoglobulin light chain genes, with the herculic goal to elucidate the complete nucleotide sequence of the entire kappa loci. Zachau and his collaborators did this work efficiently and comprehensively and, as they were the only group doing this type of work in the times before the human genome project, generated important basic information. As a result, Zachau enjoyed a considerable international reputation and was invited to many conferences to present his results. In Köhler's judgement, Zachau's application showed "little scientific phantasy" and he recommended supporting it at a much reduced level.

Other colleagues were more lucky with Köhler. When asked for suggestions for the Wilhelm von Leibnitz Prize of the Deutsche Forschungsgemeinschaft, an award connected with a research grant of considerable size, Köhler made two suggestions, Bernd Arnold and Michael Reth. Arnold, after obtaining his PhD with a thesis done at the MPIIB, had spent a successful postdoctoral period with Jacques Miller in Melbourne and had then joined the department of Günther Hämmerling at the DKFZ in Heidelberg. There Arnold continued studying immunological tolerance which he had begun with Miller. Arnold's message from many years of research was and still is that clonal deletion of T cells in the thymus is only one among many mechanisms in the induction and maintenance of self-tolerance. He derived this notion by crossing transgenic mice expressing a T cell receptor that recognized a particular MHC antigen with other transgenic mice that express this MHC antigen in various tissues. Other than normal mice which produce only a few self-reactive T cells, Arnold's double-transgenic mice produce large numbers of T cells that recognize a self-antigen, and therefore are in excessive danger to develop autoimmune diseases, unless they activate whatever mechanisms they can to tolerize their self-reactive T cells. Arnold discovered that T cells can be tolerized in a multitude of different ways, depending on many biological variables including the tissue that expresses the self-antigen, its level of expression, etc. While Arnold's work was expertly done and sound, it impressed a small ingroup in the tolerance field but did not gain the impact that it deserved in the broad immunological community. Immunologists, like most other scientists, prefer simple messages that can be explained and memorized in a few straightforward sentences. As Dimitris Kioussis once said to me, scientists are either simplifiers or complicators. Simplifiers have a much easier time getting jobs, prizes, and their papers published in high impact journals, than complicators. Arnold is a complicator. This being the reason or not, he did not get the Leibnitz Prize, in spite of Köhler's recommendation.

The person who enjoyed Köhler's support perhaps more than any other colleague was Michael Reth, who indeed got the Leibnitz Prize after Köhler's suggestion. Reth obtained his PhD with Rajewsky in Cologne making monoclonal antibodies to nitroiodophenyl (NIP) and first got to know Köhler who visited Rajewsky for advice because the hybridoma technique had stopped working in Köhler's laboratory in Basel while it was functioning well in Cologne. Reth spent his postdoctoral period with Fred Alt in New York, where he studied Abelson virus-transformed B cell tumors that resemble immature B cells arrested at various stages of heavy chain gene rearrangement, whereas the light chain genes are in an unrearranged state in most of the lines. After returning to Germany, Reth became a junior research group leader in Rajewsky's department at the Institute of Genetics in Cologne, one of the first academic environments to establish junior groups regularly and at a significant scale. There he did a widely-noted experiment showing that B cells, after rearranging their heavy chain genes once, may have a second chance of rearrangement. Before Reth's experiment the common notion was that heavy chain genes become fixed after the first successful rearrangement, presumably by some form of negative feedback of the heavy chain protein leading to allelic exclusion of heavy chain genes. Among his collection of Abelson virus-transformed B cell lines Reth found the lucky exception that was not fixed but continued in rearrangement all the way to the light chain genes. Reth looked at subclones of this particular tumor line expecting that they should have the same rearranged heavy chain gene as the parental line. To his excitement he observed new heavy chain V genes expressed in some subclones which he found to have resulted from secondary rearrangements. This was an anti-dogmatic finding as it was at variance with current concepts of allelic exclusion and suggested a novel mechanism for intraclonal somatic diversification of immunoglobulin genes. This experiment also raised Köhler's interest and he offered Reth a position in his department in the MPIIB as an independent group leader. Reth had visited the MPIIB several times accompanying Rajewsky who was in the process of negotiating his directorial offer with the MPG during 1987. Their plan was that, in case of a positive decision, Reth would join Rajewsky's department in Freiburg. As Rajewsky could not make up his mind to leave Cologne, Reth decided to join Köhler's department. Much to the dismay of Köhler the MPG did not agree to immediate tenure for Reth which Köhler had promised. Apparently the Nobel Prize-induced submissiveness towards Köhler had begun to fade. For Reth tenure was not an issue at the time, and he began working in Freiburg in 1988.

In spite of Köhler's expectations Reth soon quit studying immunoglobulin gene rearrangements and turned to the topic of B cell activation and signal transduction. Still trying to understand allelic exclusion, the first experiments Reth performed in Freiburg were transfections to express isolated heavy chain genes in his rearranging B cell line to see if the secondary rearrangements could be inhibited by expressing secreted or membrane forms of heavy chain proteins. To his disappointment, even in the presence of a proper light chain gene no immunoglobulin appeared on the surface of the transfected cells. Another transfected B cell line expressed surface immunoglobulin only on a few percent of the cells, although all of the cells received the same rearranged heavy and light chain genes. Reth isolated the population expressing surface immunoglobulin and found that these cells, in contrast to the cells without surface immunoglobulin, expressed a molecule called Mb-1 that had recently been identified in Fritz Melchers' laboratory. Transfection experiments confirmed that surface expression of the B cell receptor required Mb-1 and a second molecule of similar structure. Thus, by trying to understand allelic exclusion, Reth had discovered that the B cell receptor contained, in addition to the immunoglobulin molecule, two transmembrane molecules, nowadays termed Igα and Igβ. He had thus solved one of the pressing questions of the time, namely how the B cell receptor transmitted the activation signals which were needed to stimulate the B cell towards antibody secretion. This was a problem because the B cell receptor was known to consist of an antibody molecule that stuck with the C-terminal ends of its two heavy chains in the cell membrane but did not reach beyond the membrane into the cytoplasm of the cell. All other signalling receptors were known to either extend into the cytoplasm themselves, or to associate with other transmembrane proteins that extend into the cytoplasm, enabling them to interact with cytoplasmic molecules in charge of signal transmission, so-called second messengers. Both Igα and

Igβ reach into the cytoplasmic space of the cell, making it possible for the B cell receptor to transmit activation signals into the cell.

Reth's ultimate breakthrough to international recognition came from an observation based on amino acid sequence comparisons. When comparing the sequences of the cytoplasmic tails of Igα and Igβ with one another and with that of various different transmembrane molecules that either were themselves signalling receptors or associated with signalling receptors, he observed that they all shared several short stretches of amino acids which encompassed two tyrosines separated by two other amino acids. Other investigators had tried similar comparisons without noting the common sequence stretches. Reth's fortuitous trick was use of a color code for different groups of amino acids which made the regularly spaced tyrosines stand out to the human eye. It was already common knowledge that during cell activation tyrosines are phosphorylated and thus provide docking sites for interaction with other cytoplasmic proteins. Reth published his observation as a short letter in *Nature*, hypothesizing that the double-tyrosine stretches were important elements in the signalling function of transmembrane receptors in general. Very quickly Reth's hypothesis proved correct and the double-tyrosine stretches were for many years internationally referred to as "Reth-motif", until the name "immunoglobulin tyrosine activation motif" (ITAM) was coined. Reth's letter in *Nature*, entitled "Antigen receptor tail clue" belongs to the most frequently cited publications in immunology in recent decades.

Not much later a lucky opportunity arose for Köhler to establish Reth permanently in the MPIIB as an equal colleague. As the faculty of biology of the University of Freiburg had no department of immunology, scientists of the MPIIB covered this subject by giving lectures and courses and by supervising masters and doctoral students. These activities, which had been going on for many years, were not institutionalized or regulated by any type of formal agreement or contract. Now the faculty of biology felt that they wanted to establish a formal curriculum in immunology leading to a diploma, and approached Köhler as acting director of the MPIIB to assume the formal responsibility for teaching the entire set of courses and lectures. Köhler immediately saw the potential opportunities in this arrangement and demonstrated a considerable political talent in the way he used them to his and Reth's advantage. Responding to the request, Köhler pointed out that the main responsibility of a Max-Planck-Institute is in doing research and not in teaching. In addition, in his 3-year term as managing director of the MPIIB he would not be able to accept a long lasting responsibility for a teaching program. Moreover, he argued that with the continually changing scientific personnel of the MPIIB he could not guarantee the permanent presence of academically trained personnel licensed for teaching a formal curriculum to students. However, if the university would create a chair of immunology equipped with a basal number of positions for academic personnel, the MPIIB would be happy to provide the facilities including office space, course rooms, and laboratories for the chair, and scientists of the MPIIB would assist in the teaching program to the best of their abilities. Soon Köhler found supporters of his plan in the faculty of biology, most of all Rainer Hertel, chairman of plant genetics, who had taken a keen interest in Köhler already when he was a student (see Chapter 3), and who had written in his 1985 departmental report: "*during the period covered ... a former member of the institute received the Nobel Prize, ... and another won the world ski championship. Admittedly the honored accomplishments were achieved outside of our department, but at least they were not prevented by our influence.*" In the subsequent dealings with the University Köhler found a strong supporter in Hertel and the two developed a close friendship, a rare event in Köhler's life. The University adopted Köhler's plans and indeed succeeded in raising the finances for a new chair from the state government, certainly not an easy accomplishment. The position was then advertised and Reth was among the applicants, which included a number of potent candidates in addition to Reth. Köhler was an external member of the faculty's committee in charge of the selection, and together with Hertel managed to convince the other committee members that Reth was the proper choice for the first position on the customary list of three names. Reth took a long time deciding on the offer, and Köhler did not live to see Reth accepting the chair in 1997. For once, the negotiations between the administrations of the University and of the MPG, defining the legal status of a university chair in a Max-Planck Institute, were more complicated than Köhler and Hertel had anticipated and took a long time. In addition, Reth seemed to have learned from Rajewsky, his doctoral supervisor and long-term mentor, to stretch negotiations in order to gain any possible advantage. As a result, Reth achieved a dual appointment, chairman of immunology at the university and head of a laboratory in the MPIIB, with considerable resources from both sides. Reth about Köhler today: "*Some have criticized him that he was not always aware of the current state of knowledge in his field, and often was not with the cutting edge of the scientific progress. However, Georges could look at a scientific problem from an unusual angle, and could open your eyes to a point of view that you would never have seen yourself. He was a player, he played with science and for him science was a kind of play.*"

# Chapter 13
# Post-Nobel science I

During his last year in Basel Köhler took a decision which later turned out to be nearly as fortuitous as that of making hybridomas. It was based on a hunch that the foreseeable future of immunology was going to be based on genetic manipulation of mice, a shot that could not have been closer to the mark. As a result, he set out to produce transgenic mice by implanting a pair of rearranged antibody heavy and light chain genes, in order to make them express the same antibody molecule in all their B cells. There were several pressing questions in immunology that one could hope to answer with transgenic mice expressing single antibody molecules. One problem was how allelic exclusion was established. Cotton and Milstein were the first to show that in hybrid cells obtained by fusing a mouse and a rat myeloma both parental antibody heavy and light chains were produced, leading to co-expression of the two parental antibodies and of a multitude of mixed forms in the same cell. They had thus demonstrated that a single cell can survive making more than one antibody. Nevertheless, each normal B cell only expresses a single antibody, consisting of one heavy and one light chain, although it has two sets of chromosomes to make at least two, possibly more, rearranged heavy and light chain genes. Does a normal B cell avoid making more than one set of rearrangements, or do normal B cells with more than one set of rearrangements die, in spite of what Cotton and Milstein found for malignant myeloma cells? Since all B cells in a transgenic mouse would possess the same pair of rearranged antibody transgenes, an analysis of the rearrangements of the endogenous antibody genes could perhaps answer this question. There was also the question of self-nonself discrimination. How is self-tolerance established, i.e., how does an animal deal with B cells that produce antibodies against self-antigens? If one could make transgenic mice expressing a self-reactive antibody, one could perhaps study this situation. While these future questions were in the back of the minds of the scientists, they had first to solve a myriad of technical problems of the new endeavor.

The first problem was how to isolate rearranged antibody genes for transgenesis. Normal B cells were not a promising source, as each would have another pair of heavy and light chain genes rearranged so that one would isolate a mixture of many different antibody genes. In contrast, all cells of a hybridoma or a myeloma express the same pair of rearranged heavy and light chain genes, and investigators had the choice between hybridomas and myelomas to isolate individual rearranged antibody genes. Several investigators, including David Baltimore at Massachusetts Institute of Technology, Ursula Storb at Harvard University, and Köhler in Basel, began to isolate antibody genes for making transgenic mice. Baltimore's group chose a myeloma with the notorious lack of specificity which, however, was irrelevant for the question of allelic exclusion. Alternatively, the antibody genes could be isolated from hybridomas with known specificities. This choice seemed much preferable to Storb and to Köhler, as the mice could be useful for studies that depended on antibody specificity. Storb used a hybridoma that had been made in Rajewsky's laboratory and produced antibodies to NIP, whereas Köhler used his SP6 hybridoma with specificity for TNP (see Chapter 6). For isolation of the antibody genes Köhler had the help of colleagues at the BII and elsewhere. They first made a preparation of DNA from lysates of SP6 cells. They then fragmented the long strands of DNA into shorter pieces using restriction endonucleases which cut DNA at certain short nucleotide sequences that are irregularly spaced in the DNA so that pieces of varying length are obtained. The DNA fragments were then randomly ligated into DNA of λ phage, a bacteriophage that infects the bacteria, multiplies and eventually lyses them. The infected bacteria were then plated on agar where the phage grew as clones by lysis and reinfection. In order to find the phage clones that contained the rearranged heavy and light chain genes, short copy-DNA probes homologous to the constant regions were prepared, made radioactive, and then hybridized to a blot taken from the agar plate on which the phage clones were growing. Autoradiography of the blot then showed a black spot in the position of each phage clone containing a heavy or light chain gene, depending on the probe used for hybridization. After growing the identified phage clones to large numbers, Köhler and his colleagues could recover large amounts of λ phage DNA containing the isolated heavy and light chain genes of SP6. The two genes were then recombined by cloning them in tandem into a single plasmid that was suitable for making transgenic mice. For this purpose Köhler collaborated with Sandro Rusconi in Zürich who mastered the method of DNA injection into the mouse zygote. After fertilization of the oocyte by a sperm and before nuclear fusion, the young zygote contains for several hours two separate pronuclei, the male and the female pronucleus. These zygotes are obtained by flushing the uterus of the female after carefully timed fertilization. For injection of DNA one uses a micromanipulation device which allows fixation of the zygote to the tip of a blunt hollow needle by gentle suction. A sharp needle is used to inject the

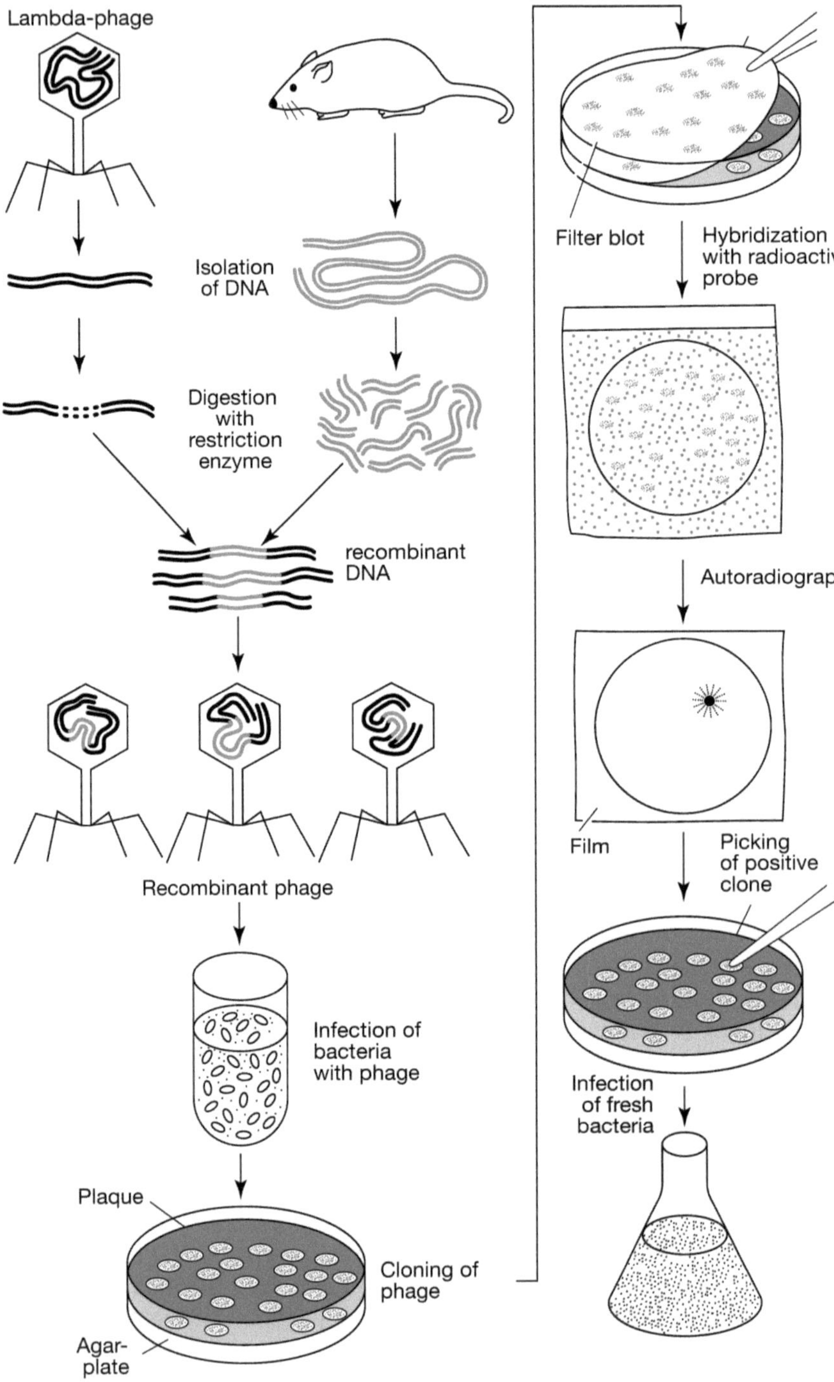
Lambda-phage
Isolation of DNA
Digestion with restriction enzyme
recombinant DNA
Recombinant phage
Infection of bacteria with phage
Plaque
Cloning of phage
Agar-plate
Filter blot
Hybridization with radioactive probe
Autoradiograph
Film
Picking of positive clone
Infection of fresh bacteria

plasmid DNA under microscopic control into the male pronucleus, which is the bigger of the two and therefore easier to hit. Rusconi injected between 20 and 100 copies of the plasmid prepared by Köhler into each zygote. If all goes well, some of the injected DNA inserts itself into one of the chromosomes of the male pronucleus by a seemingly random process, integrates into the genome of the resulting embryo and is transmitted to its offspring, thus giving rise to a transgenic mouse line. 13 zygotes were injected by Rusconi, and five mice were born, all containing the transgenes in their genome.

While Storb's and Baltimore's initial mice were made with only a heavy or only a light chain gene, respectively, Köhler and Rusconi were the first to generate a transgenic mouse using a pair of heavy and light chain genes. The first results were highly promising. First of all, the antibody transgenes were expressed in essentially all of the B cells but not in other tissues of the transgenic mice. This meant that Köhler's plasmids contained all genetic control elements that are required for expression of antibody genes specifically in B cells and nowhere else. When Köhler isolated his antibody genes from SP6 hybridoma cells the control elements, termed promoters and enhancers, for immunoglobulin genes were not completely known and he was lucky that the DNA pieces that he isolated contained these essential sequences. Morover, a major proportion of the serum immunoglobulin was the product of the transgenes, suggesting that the transgenes induced the phenomenon of allelic exclusion. Thus, B cells of transgenic mice obeyed the rule of normal B cells each of which expresses only a single pair of antibody genes. An analysis of the rearrangement of endogenous antibody genes revealed that the heavy chain genes had remained either unrearranged or stopped rearrangement after D to J joining without being joined by a V-gene. In contrast, endogenous light chain genes appeared to rearrange normally. The data did not support a putative elimination by cell death of double expressing B cells but rather suggested a feedback mecha-

---

Figure 27
Procedure for cloning mouse genes in bacteria. DNA is isolated from a mouse organ or from a mouse cell line such as a hybridoma. The DNA is cleaved into fragments by restriction enzyme digestion and mixed with the DNA of λ bacteriophage digested by the same enzyme. The mouse DNA fragments recombine with the phage DNA by replacing the center fragment which is not essential for the phage. The recombinant phage are then mixed with *E. coli* bacteria so as to infect a small proportion of them. The bacteria are plated on agar as a confluent lawn. Infected bacteria get lysed, producing new phage which lyse neighboring bacteria so that clear spots are formed in the bacterial lawn, each by a clone of phage containing a mouse DNA fragment. A filter is then placed on the agar, blotting up the DNA as an exact replica of the agar plate. The filter is then lifted and hybridized with a radioactive DNA copy of the mRNA corresponding to the gene one wants to isolate. Subsequent autoradiography of the filter reveals the position of phage clones that contain the DNA fragments in question. The phage clones thus identified can then be recovered from the agar plate and augmented in new bacteria to yield large amounts of recombinant DNA containing the gene in question (from Leder, 1982).

nism exerted by the first rearranged antibody genes, preventing further rearrangements of heavy chain genes but not of light chain genes in the same B cell. In transgenic B cells, these are the transgenic heavy and light chain genes because they are present in rearranged state already in the genome and therefore earlier than all others. As a consequence, the majority of B cells of such transgenic mice express the transgenic antibody while only a few B cells make endogenous antibodies. In other words, mice expressing a pair of antibody heavy and light chain transgenes are monoclonal mice, their B cells produce predominantly one antibody with one specificity. The antibody produced by the SP6 transgenic mouse belonged to the IgM class, i.e., it possessed a μ heavy chain paired with a κ light chain, and specifically recognized TNP.

But how does a pair of rearranged antibody genes exert a negative feedback on subsequent heavy chain rearrangements in the same cell? In the mid 1980s a host of knowledge about the genetics, biosynthesis, and regulation of the antibody system had accumulated. Antibodies of the IgM class, the first class to be produced in an immune response, form pentamers consisting of five typical monomeric molecules composed of two μ heavy chains and two light chains, either of κ or λ type. Naive B cells, i.e., B cells prior to their first contact with an antigen, possess two immunoglobulin classes as surface receptors, IgM and IgD, containing μ and δ heavy chains, repectively. Following initial contact with the antigen, B cells start to secrete their IgM, whereas IgD is never secreted. Upon repeated exposure to the same antigen the B cells, now called memory B cells, shut down the synthesis of IgM and IgD and switch to the IgG classes or in rare cases to IgA or IgE. Similar to IgM on naive B cells, IgG first appears as receptor on the memory B cell surface and is then secreted after the B cell has been restimulated by antigen contact. IgM displays 10 identical combining sites and therefore a low binding affinity suffices for successful attachment of an antigen. IgG is a monomer and thus has only two antigen attachment sites. This requires a higher binding affinity for successful antigen binding. The high affinity is acquired by random somatic hypermutation of antibody V genes followed by selection of high-affinity mutants (see Chapter 7). On the genomic level the immunoglobulin gene rearrangement and class switch is achieved by a complicated but highly regulated cascade of recombination events. It starts in immature B cells with transcription of the unrearranged immunoglobulin heavy chain locus, which is regulated by promoter/enhancer elements in the intron between the J cluster and the Cμ gene. This so-called "germline transcription" makes the locus accessible to DNA recombinases, termed Rag1 and Rag2, which are responsible for cutting and religating the DNA, giving rise first to a DJ and in a second step to a complete VDJ rearrangement. The VDJ gene comes to sit in front of the Cμ gene that encodes the constant region of the IgM heavy chain, separated from it by an intron. The Cδ gene that encodes the constant region of the heavy chain of IgD is next down the line, followed by several Cγ genes for the subclasses of IgG heavy chains, a Cε gene for the IgE heavy chain, and a Cα gene for the IgA heavy chain, all separated by introns. In naïve B cells a long RNA transcript is made spanning the VDJ exon, Cμ and Cδ, and including the two introns in between. This long transcript simultaneously serves to synthesize both μ and δ heavy chain mRNAs and proteins, by splicing the VDJ sequence either to Cμ or to Cδ respectively. In memory B cells, a DNA segment encompassing Cμ and Cδ plus a varying number of the downstream C genes is cut out so that the VDJ gene now sits in front of the next Cγ gene down the line, or in front of Cα or Cε. A new RNA transcript is made including the VDJ segment, the intron, and the following C gene, and mRNA is produced by splicing the VDJ sequence of the transcript to the C region. The memory B cell, that has now lost the ability to produce IgM and IgD, makes a new class of antibody which is determined by the heavy chain protein encoded by its new C gene. The VDJ segment responsible for the antigen binding specificity, as well as the light chain, have remained the same, but may have acquired a number of mutations.

While all of this was known for some time, the knowledge did not lead to a readily testable hypothesis to how allelic exclusion was achieved. Meanwhile in Freiburg, Köhler considered the possibility that the answer to the question could be found in the regulation of germline transcription, as it opened up the heavy chain locus to recombinase activity. To this end, he asked his new PhD student, Andreas Kottmann, to search for additional enhancer/promoter elements. Kottmann, a very promising young investigator, did a series of state of the art experiments which resulted in the discovery of additional enhancer/promoter sequences in the region around the most downstream D segment, termed DQ52. Kottmann, Köhler's favorite student, represented his work at a site visit of the scientific advisory board of the MPIIB and received the highest marks in the report. However, a link of Kottmann's results to allelic exclusion could not be made. The second approach that Köhler took was to ask if heavy chain classes other than the μ chain could also exert the negative feedback that terminated the rearrangement process. As IgM and IgD are expressed simultaneously in naive B cells, it seemed reasonable to ask if a rearranged δ transgene had the same effect as a rearranged μ transgene. In addition, as nobody had a clue about the functional role of IgD in immune responses, Köhler hoped that an IgD-transgenic mouse would perhaps yield information on the function of the only non-secreted immunoglobulin class. Köhler's collaborators, Antonio Iglesias and Marinus Lamers, made a new transgenic mouse using a construct in which the Cμ gene was deleted and the heavy chain Vgene of SP6 was inserted directly in front of Cδ, so that only transgenic δ heavy chains could be produced. Unexpectedly, B cells of the transgenic mouse secreted IgD into the serum, a finding that could not be easily explained. However, the analysis of the rearrangement of endogenous immunoglobulin genes yielded a clearcut result. Similar to the IgM-transgenic mouse, most endogenous heavy chain genes remained unrearranged or were arrested after D to J rearrangement. The data clearly showed that both rearranged μ and δ heavy chain genes could inhibit further rearrangements. However, neither the physiological function of IgD nor the feedback inhibition of rearrangement could be elucidated by Köhler's transgenic mice. Both questions turned out to be tough ones and are still not completely answered today. As far as allelic exclusion is concerned, due to the studies of Melchers, Rajewsky, Micheal Nussenzweig, Fred Alt, and others, we know that in immature B cells the immunoglobulin heavy chain is exposed on the cell surface together with a so-called surrogate light chain, consisting of a pair of invariant molecules. This pre-B cell receptor is associated with the signalling molecules Igα and Igβ and is required for further maturation and induction of light chain rearrangement, resulting in mature B cells expressing a complete B cell receptor consisting of heavy chains and regular light chains. In the meantime it had been learnt that a μ chain without a regular light chain was sufficient for allelic exclusion, and it seemed obvious that the pre-B cell receptor was responsible for allelic exclusion of the heavy chain genes. This hypothesis was attractive to many experts in the field, until Rajewsky put it to the test by generating mice in which the surrogate light chain was deleted. As a surprise, allelic exclusion of heavy chain genes was not strongly impaired in these mice, although they were unable to produce a complete pre-B cell receptor. Similarly, Lamers later showed that mice in which IgD is genetically deleted show no obvious immune dysfunction.

Köhler's SP6 transgenic mice were useful for some less spectacular studies that former colleagues in Basel performed in collaboration with Köhler, who generously provided the mice to interested colleagues. For example, the fact that the majority of B cells had TNP-specific immunoglobulin receptors provided a unique opportunity to force them to interact with T cells. The interaction was mediated by adding a monoclonal antibody that bound to the T cell receptor and had been coupled with TNP to be recognized by the receptors of the SP6 B cells. Adding this reagent to a culture of T cells and SP6 transgenic B cells or injecting it into SP6 transgenic mice resulted in the formation of a physical bridge between receptors of the TNP-specific B cells and of T cells. In one study using this antibody reagent, SP6 transgenic B cells were forced to interact with T helper cells, and the authors could show that the B cells could indeed receive helper signals by this type of artificial interaction, although in the normal situation the T helper cell recognizes a peptide displayed by the MHC class II molecules on the B cell. In another study, SP6 B cells were forced to interact with killer T cells. The authors showed that the B cells were indeed killed by this interaction, although in the normal case the T cell receptor of a killer cell recognizes a peptide displayed by the MHC class I molecules on target cells. The results were of interest but did not have a lasting impact on concepts of T cell help or cytotoxicity. They are mentioned here because they represent a significant proportion of Köhler's published work in 1986 and 1987.

Unfortunately, the SP6 transgenic mice did not lend themselves for studies on tolerance, as TNP is not a self antigen and, being a small chemical, cannot be expressed as a self-antigen by gene transfer in transgenic mice. For attacking the tolerance problem, Köhler and his collaborators had to make new transgenic mice. One of their strategies was based on the availability of monoclonal antibodies recognizing antigens on the surface of lymphocytes. Such monoclonal antibodies had meanwhile been produced in large numbers by various research laboratories and companies for the typing of mouse and human lymphocyte subsets in research and medical diagnosis. In the mouse, the antigens recognized by such monoclonal antibodies can be taken as self-antigens and Köhler, together with his collaborator, Hermann Eibel, and a PhD student, Frank Brombacher, focussed on CD8, a protein expressed on killer T cells and on certain subsets of thymocytes. In order to make a transgenic mouse expressing an antibody to this self-antigen, they used a hybridoma secreting antibodies to CD8 as source of antibody genes. Unfortunately, Brombacher, Eibel, and Köhler took a convenient shortcut strategy which later turned out to be a serious drawback for the project: They isolated only the heavy chain gene, relying on their own experience and that of others that transgenic heavy chains usually find a sufficient number of endogenous light chains for restoring the antigen binding specificity of the original heavy chain/light chain combination. The phenotype of this transgenic mouse consisted primarily of transgenic artifacts. First, while the transgenic heavy chain was expressed, they could not find CD8-specific antibodies in the transgenic mice, a result that was interpreted as deletion of the B cells in which the transgenic heavy chain had paired with suitable light chains to produce anti-CD8 antibodies. However, due to the lack of a transgenic light chain as a marker for the anti-CD8 specificity they could provide only indirect proof for this speculation. Second, the mice showed a drastic reduction in thymocyte number. Because no transgenic anti-CD8 antibodies were found, the deletion of thymocytes could not be caused by circulating anti-CD8 autoantibodies. Rather, and again unexpectedly, Eibel and Köhler found that the transgenic heavy chain was expressed intracellularly in thymocytes. The transgene thus did not obey the rule of tissue-specific expression only in B cells, there was something wrong with its control elements. They speculated that the heavy chain, in the absence of a light chain, would combine with the CD8 protein synthesized in the tymocytes, and that this interaction would arrest thymic differentiation. Again, there was no proof for this hypothesis. Owing to Köhler's reputation, they managed to publish this inconclusive work, representing the major result of the department in 1989/1990, in the *EMBO Journal*, the official journal of the European Molecular Biology Organization, and in the *Journal of Experimental Medicine*, both in the upper ranges of the impact scale. Other investigators would not have got these papers through the peer review process, or would perhaps not have attempted to publish these results in the first place.

Köhler produced two other transgenic mice with the aim to study tolerance and autoimmunity. In one of them, the idea was to introduce a foreign gene into the mouse so that the molecule encoded by the gene would be expressed in the mouse as an autoantigen. One could then compare the immune responses to the molecule in normal and transgenic mice and, hopefully, find qualitative or quantitative differences. Köhler chose β-galactosidase, the bacterial enzyme that he knew well from his doctoral thesis in which he had studied immune responses to this antigen in several species including mice. The technique of generating transgenic mice had meanwhile been established as routine in Köhler's laboratory so that he could ask a PhD student to generate a proper DNA construct for injection into zygotes. Probably out of laziness they used the μ chain construct previously employed for making the SP6 transgenic mice, and just replaced the μ gene by the β-gal gene so that β-galactosidase was expressed under the control elements of the immunoglobulin heavy chain locus, i.e., in B cells. Köhler probably thought that expression of an autoantigen in B cells is as good as in any other tissue, but others might have guessed that expression of the autoantigen in B cells creates a complicated situation compared to other tissues, as B cells are the very cells that are subject to becoming tolerized as well. Whether this was the reason or not, the results were hard to interprete. When immunized with β-galactosidase the transgenic mice indeed produced less antibodies than normal control mice. Immunization with bovine serum albumin as a contol antigen showed no difference in antibody levels. While these observations seemed to support the conclusions that at least a partial tolerance to β-galactosidase had been induced by the transgene, the problem started when they immunized the mice with a covalent complex of β-galactosidase and bovine serum albumin. In this case, the antibody response to bovine serum albumin was also reduced in the transgenic mice. As this could not readily be explained by β-galactosidase-specific tolerance of either T cells or B cells, Köhler suggested a complicated dominant suppression mechanism induced by B cells presenting β-galactosidase, as well as structures linked to β-galactosidase, to T cells which would then suppress the B cell response. The mechanism remained in the field of speculation, and thus had no impact on current concepts of tolerance. Using a more carefully designed and more elaborate transgenic system, Cris Goodnow, a young Australian working in the laboratory of Tony Basten in Melbourne, had published very clear and convincing results on B cell tolerance 2 years earlier. Goodnow had produced two transgenic mice. One expressed hen egg lysozyme as an autoantigen under the control of the metallothionin promoter, so that the transgene was not expressed in untreated mice but its expression could be induced at any given time point in all tissues by injection of zinc ions into the mouse. The other transgenic mouse expressed a monoclonal antibody to hen egg lysozyme. Mice obtained by crossing the two transgenic mice now expressed both the autoantigen and the antibody recognizing it, the latter on the majority of the B cells. Goodnow found that the B cells were not eliminated but continued to exist following induction of hen egg lysozyme. However, they became inactivated, i.e., they could not be activated by the autoantigen to secrete antibodies. These influential experiments defined clonal anergy as a major mechanism of self-tolerance in B cells. Another set of well designed tolerance experiments using transgenic mice was done in the BII, certainly with Köhler's knowledge, and also published before Köhler's work. David Nemazee and Kurt Bürki at the BII had produced a transgenic mouse expressing a monoclonal antibody to a particular MHC class I molecule. When this mouse was crossed with mice of the corresponding MHC type, B cells expressing the transgenic antibody were eliminated. These results demonstrated that the immune system can also use clonal deletion of autoreactive B cells as a means to induce self-tolerance. Later studies of these and other investigators showed that both mechanisms of B cell tolerance coexist in the immune system.

In designing their third transgenic mouse, Köhler and his collaborators joined a club of several groups of investigators trying to construct mouse models for autoimmune diabetes. This disease is associated with the failure of the β-cells in pancreatic islets to produce insulin. In humans suffering from this form of diabetes, also referred to as type I or juvenile diabetes, one observes expression of MHC class II molecules on the β-cells which are not seen in healthy controls, and the islets are infiltrated by T lymphocytes. Together this suggests that autoimmune T lymphocytes attack the β-cells and destroy them. To see if the aberrant expression of MHC antigens on the β-cells is responsible for the disease, transgenic mice were constructed independently in several laboratories in which MHC transgenes were expressed under the control of the insulin promoter, using different class I or II genes. The results obtained by most of the groups revealed that the mice developed diabetes but did not show the infiltration of the pancreatic islets with T lymphocytes. This suggested that the impairment of insulin production in the transgenic mice was not due to autoimmune destruction of β-cells as in the human disease, but to some other effect of transgene expression in these cells. This was most likely the competition for transcription factors that bind to the insulin promoter in order for the insulin gene to be transcribed. In normal β-cells there are only two copies of the insulin promoter whereas in transgenic β-cells there are multiple additional copies of it, one with each copy of the transgene. It is therefore possible that the amount of transcription factor is limiting leading to reduced availability for the insulin promoter and impaired insulin production. Because it was clear to the investigators that they had generated a "transgenic artifact", the real reasons for the impaired insulin production were not very interesting and therefore not elucidated. However, it was generally agreed that this type of transgenic mouse was not a suitable model of autoimmune diabetes. Most of the other groups had these results long before Köhler, had presented them at conferences and published them. Köhler nevertheless continued producing these mice and, to no one's surprise, obtained similar results.

# Chapter 14
# Post-Nobel science II

Moreso than transgenic mice, Köhler correctly foresaw that gene knockout mice were to become the key source of scientific progress in immunology in the last decade of the 20th century. While in transgenic mice one or several new genes are artificially added to the mouse genome, in knockout mice one or several genes are destroyed. Depending on the importance of a gene for the life of the mouse organism, its destruction may be incompatible with life so that the mouse embryo dies at an early stage before birth. This is the case for the essential genes, including many developmental and the so-called housekeeping genes, whose destruction does not lead to living offspring and whose analysis by the knockout technology could establish merely the fact of lethality. However, the mammalian organism contains many genes that are not essential for life, and whose destruction therefore allows the birth and growth to adulthood of mutant individuals. These so-called knockout mice show more or less severe anatomical or functional deficiencies, the so-called "phenotype", the characterization of which provides information on the function of the destroyed gene. The genes of the immune system, although essential for life in a normal environment full of microorganisms, belong to these non-essential genes, because mice with an impaired immune system can be readily kept healthy in the controlled microbial challenge of a laboratory. The genes of the immune system and the molecules they encode were therefore particularly amenable to analyses by the knockout technology.

Making knockout mice requires the mastering of at least three highly demanding technical challenges, the culture of embryonic stem (ES) cells, the exchange of genes in ES cells by homologous recombination, and the production of chimeric mice by injection of ES cells into mouse blastocyst stage embryos. Embryonic stem cells are cell culture lines derived from the inner cell mass of a mouse blastocyst, a 64-cell stage mouse embryo just before implantation into the uterus mucosa. The blastocyst already consists of two types of tissue, the trophoblast, a hollow sack which later develops into the placenta, and the inner cell mass which later develops into the embryo proper. This split into two different cell types is the first apparent differentiation event in embryogenesis, and is followed by a multitude of further differentiation steps which finally lead to all the different organs and tissues of the mature mammalian organism. In order to study early differentiation, developmental biologists tried for a long time to obtain cell lines of embryonic origin that would perform at least some of these differentiation steps in tissue culture. The first such cell lines that became available were the so-called teratocarcinoma cells. Some teratocarcinomas, malignant tumors that arise in mice and humans and consist of a mixture of undifferentiated and various differentiated tissues, have been adapted to tissue culture lines with some capacity of differentiation. After initial excitement, however, researchers realized that differentiation of teratocarcinoma lines was rather minimal, and started to look for alternatives, if possible of non-malignant origin. Martin Evans and Matthew Kaufman at the University of Cambridge in England were the first to report the successful establishment of a tissue culture line derived from the inner cell mass of mouse blastocysts in 1981. It took them a long time to figure out the conditions for continuous growth of these cells in culture and they came up with a complicated protocol. First, one had to use delayed implantation blastocysts derived from ovarectomized females treated with progesterone to prevent implantation. Under these conditions the blastocysts grow to larger size and cell numbers. Moreover, one had to use a feeder cell layer and they recommended a particular transformed embryonic cell line that grew continuously and was therefore difficult to get rid of again.

In spite of the cumbersome protocol, Evans's and Kaufman's success seemed a real breakthrough and a number of investigators working in embryology tried to repeat their experiment to establish their own ES cell lines. One of the first who succeeded was Rolf Kemler, at the time junior research group leader in the Friedrich-Miescher-Institute, a junior group laboratory associated with the Max-Planck-Institute of Developmental Biology in Tübingen. Kemler, a veterinarian by university training, had obtained a PhD degree with a thesis done at the MPIIB in Freiburg. In the same age group as Köhler, Kemler got to know Köhler when both took PhD courses organized by Fritz Melchers at the University of Freiburg. Kemler then went to join Francois Jacob for a postdoctoral period at the Pasteur Institute in Paris where he initially studied teratocarcinoma cells but soon became frustrated with them. Kemler then turned to study cell adhesion mechanisms in the developing embryo and discovered a molecule now called E-cadherin, then dubbed uvomorulin, which he found to be responsible for the tight adhesion among the cells of embryos at the morula stage, a process termed compaction. With these widely noticed results Kemler was appointed as head of one of the four highly sought-after Miescher Units in Tübingen, at the time considered a sure start for a first rate academic career. Kemler continued studying E-cadherin in Tübingen, but also became intrigued by the results of Evans and began culturing embronic stem cells. He was the first in Germany to succeed and significantly improved the technology by making it simpler. He found that one could use normal blastocysts instead of delayed implantation blastocysts, and introduced the use of embryonic fibroblasts as feeder cells, non-transformed primary cells with a limited lifespan which are easily derived from dissected mouse embryos. The ES cell line D3, derived by Kemler in Tübingen, was distributed all over the world and was for many years the ES cell of choice in the production of knockout mice.

The group in Cambridge including, besides Evans and Kaufman, Allan Bradley and Elizabeth Robertson, demonstrated in 1984 that cultured ES cells could be reinjected into a new mouse blastocyst and that these cells would become part of the inner cell mass giving rise to a chimeric embryo in which some of the tissues would be derived from the injected ES cells, whereas others would be derived from the inner cell mass of the original blastocyst. In some cases, the injected ES cells would even give rise to germ cells so that their genes would be transmitted to the offspring of the chimeric mouse. ES cell lines able to integrate into the germline were called "germ line competent" and opened a novel and fascinating way of manipulating the mouse genome. The Cambridge group was the first to show that this was possible by infecting ES cells with a retrovirus, and demonstrating genetic transmission of retroviral genes to the offspring of chimeric mice. However, not all ES cell lines were germ line competent and many ES cells lost germ line competence after a few passages in culture, indicating that maintaining germ line competence required very careful culture procedures. Manipulating the genome of ES cells, however, often requires rather rough treatment of the cells. This is particularly the case if one needs to use antibiotic selection to isolate the few ES cells in which the manipulation was successful. The main approach in use to select genetically manipulated mammalian cells was to introduce a bacterial gene, the neomycin resistance gene, together with the plasmid used in the manipulation. Cells which had integrated the plasmid into their genome were then selected by treatment of the culture with neomycin, which eliminated all cells that had failed to integrate the plasmid. Would it be possible to maintain germ line competence in ES cells transfected with the neomycin resistance gene and selected by treatment with neomycin? Kemler did this experiment with his D3 cell line and constructed a strain of mice in which the neomycin resistance gene was genetically transmitted. This experiment, published in 1986, showed that germ line competence of ES cells could survive neomycin selection and thus was a milestone on the way to the knockout technology.

Introducing a novel gene into ES cells and channelling it into the mouse germline by injecting the ES cells into blastocyts was a technical advance but otherwise represented just a more cumbersome way of making transgenic mice, which can be made more easily by injecting DNA directly into the male pronucleus of the zygote. The purpose of doing these preparatory experiments was a different, more sophisticated type of genetic manipulation, namely the exchange of genes by homologous recombination. As previously shown in bacteria, by homologous recombination it is possible to replace one or several adjacent genes by other DNA of one's own choice. To do so, a piece of DNA is islolated from the bacterial genome which is complementary to the DNA of the targeted gene and includes its flanking regions on both sides. By standard recombinant technology the targeted gene is modified, for example by introducing some form of mutation that inactivates the gene. The artificial DNA, termed the "construct", is then introduced into the bacteria to recombine with the bacterial genome. The process of recombination is to a large extent random, so that the introduced DNA statistically inserts itself at any position into the bacterial DNA. However, recombination is facilitated by complementarity, so that recombination with the homologous DNA at the complementary position of the bacterial genome occurs with a greater than random frequency. The longer the flanking complementary regions in the construct, the higher the probability of homologous recombination. If homologous recombination occurs such that both ends of the construct recombine to the complementary DNA, the wildtype gene is replaced by the mutant gene in the construct. The bacteria thus carry a mutant gene instead of the wildtype gene, and the effect of the mutation on the bacterial phenotype can be studied.

If genes in ES cells could be replaced by homologous recombination, it would be possible to generate mouse strains that carry mutated genes just like bacteria. The idea fascinated biologists, but for some time it seemed doubtful that it could ever work. Because the genome of mammalian cells is ultimately more complex than that of bacteria, containing long introns and lots of repetitive sequences, it was not clear if a construct would ever find its homologous DNA, not to speak of recombining to it. Moreover, it seemed hopeless to identify the few ES cells in which homologous recombination had taken place. While random integration of introduced DNA can be expected to take place in several percent of the cells, homologous recombination was estimated to occur orders of magnitude less frequent, if at all. In the end, Oliver Smithies at the University of Wisconsin and Mario Capecchi at the University of Utah convinced the scientific community that homologous recombination indeed took place in mammalian cells, one only needed very efficient methods of selection to recover the few cells in which homologous recombination had occurred. Both investigators independently succeeded to mutate the hypoxanthine-guanine phosphoribosyltransferase (HGPRT) gene so that the ES cells could be selected with 5-azaguanine (see Chapter 5). To select homologous recombination events of any other targeted gene in ES cells, Capecchi introduced the so-called positive–negative selection scheme, which combined two selection principles. Positive selection employed the well-known neomycine resistance gene, which was recombined somewhere into the sequence of the targeted gene of the construct, thus serving the purpose of disrupting the targeted gene at the same time. Selection with neomycin, however, was insufficient, as all cells that had integrated the construct, either by random integration or by homologous recombination, expressed the neomycin resistance gene and hence would survive selection. Therefore a second, negative selection step was necessary. It utilized the thymidine kinase gene of herpes simplex virus which had to be attached to one end of the construct outside of the region of complementarity. Viral thymidine kinase, in contrast to mammalian thymidine kinase, phosphorylates the nucleoside analog Gancyclovir into a toxic nucleotide which is incorporated into the DNA and kills the cell. In most random integrations of the construct the herpes simplex virus thymidine kinase gene would be included, so that these cells would be killed upon culture in the presence of Gancyclovir. In contrast, in most homologous recombination events the herpes simplex virus thymidine kinase gene would be deleted so that these cells would survive culture in Gancyclovir. In principle, ES cells that have integrated the construct by homologous recombination would therefore be the only cells surviving double selection. With this selection method all technical steps were at hand to generate knockout mice for any known gene of anyone's interest. The first knockout of a gene of interest in immunology was done by Rudolf Jaenisch and coworkers. They destroyed the gene for β2-microglobulin, a protein required for expression of MHC class I antigens at the cell surface. The mouse had a drastic phenotype, lacking all killer T cells owing to the missing MHC class I expression in the thymus. This first immunological knockout mouse was the beginning of a success story of the knockout technology in the genetic unravelling of the immune system.

Köhler wisely foresaw the potency of the knockout technology and took measures to establish it in the MPIIB. Kemler's 5-year period as head of a Miescher group was running out in 1987 and, being married to a French woman and French being the first language of their children, planned to go back to France. An option at the Institute Gustave Roussy in Villejuif was cancelled at short notice for political reasons, and so Kemler was looking for new opportunities. Köhler saw his chance to hire one of the most experienced experts in the ES cell field, and offered Kemler a position as independent group leader in the MPIIB. Kemler was somewhat hesitant for several reasons. First of all, he was afraid that he would miss the intellectual interaction with other developmental biologists. Immunology was the only subject of study at the MPIIB at the time and Kemler had always taken a dim view of this field. He had adopted an attitude which used to be trendy among biologists, namely that immunology was beyond comprehension of any self-respecting scientist. Moreover, Köhler had offered him laboratory facilities in the oldest part of the institute, of low technical standard and small in addition. Nevertheless, the foundation pits of the new laboratory building had already been dug out, and Köhler promised Kemler laboratories there once it had been completed in an anticipated period of about 2 years. In the end, Kemler had no other job offer and thus moved to Freiburg in the fall of 1987. There he soon found his poor expectations confirmed, his laboratories were of a quality that was incompatible with the high standard required for ES cell culture. Kemler thus dropped working on ES cells for the time being and concentrated on his other field of interest, the study of E-cadherin. Here he soon made several other important contributions, on the basis of which he was appointed director at the MPIIB in 1992. Most importantly, he discovered the catenins, a group of three proteins that associate with the cytoplasmic tail of E-cadherin and regulate its adhesion functions. Among those, β-catenin soon became a focus of worldwide interest, as it turned out to be a central factor in a multimolecular control circuit regulating cell physiology and homeostasis. In addition to its involvement in cell adhesion, β-catenin can act as transcription factor. Above a certain concentration in the cytoplasm β-catenin is translocated into the nucleus and stimulates transcription of a number of target genes. The cytoplasmic levels of β-catenin are controlled by Wnt signalling, a highly conserved pathway of signal transduction first described in *Drosophila*. In mammals, Wnt signalling regulates β-catenin levels through phosphorylation which facilitates protein degradation by the ubiquitin–proteasome system. Deregulation of this control circuit results in increased β-catenin levels and malignant transformation of the cell, leading to intestinal polyposis and carcinoma of the colon.

Kemler's initially poor laboratory facilities were not the only factor that slowed down the establishment of homologous recombination in ES cells at the MPIIB. Kemler today: "*Georges never indicated to me that he wanted me to solve his scientific questions. I never had to defend myself against impositions of that kind. He was interested in the technology and it was our understanding that I help establish it in his own department. However, there were quite a number of problems in his department. George advanced from a postdoc situation in Basel, where he just had to reach into a refrigerator to find all reagents, to director of a department. These were too many steps taken at once. As a consequence, there were management problems resulting in a degree of unrest in his department. For example, of the four group leaders he hired two got permanent contracts and two did not. This created jealousy and hampered the cooperation among the group leaders. In addition, the person he asked to be in charge of setting up the ES cell technology was incompatible, I knew this within the first half hour of meeting him.*"

All of this resulted in a considerable delay in the ability of Köhler's department to generate knockout mice. In the German national competition, he was beaten by a clear margin by Rajewsky, who generated the knockout mouse that Köhler might have been aiming at, a mouse unable to produce B cells. Using homologous recombination in ES cells, Rajewsky deleted the membrane exon of the Cμ gene, so that only the secreted but not the membrane form of the μ chain could be produced. As a result, no pre-B cell receptor could be expressed at the cell surface which requires in addition to the surrogate light chain the membrane form of the μ chain.

Since B cell development depends on signals through the pre-B cell receptor, no mature B cells are produced in this knockout mouse. Rajewsky's paper describing his first knockout mouse was published in 1991. Subsequently the mice were made available to interested colleagues and became an important tool for studies on the role of B cells in the immune system. In the years to follow Rajewsky deleted a large number of genes, including many genes that were suspected to have a role in B cell development and function, and thus cemented the position of his laboratory among the two or three leading ones in the B cell field. Moreover, his laboratory took a key role in making further technical advances to the knockout technology.

Köhler had hired a new student, Manfred Kopf, who turned out to be a lucky choice and became the driving force in Köhler's knockout program. Kopf, who is now professor at the Swiss Federal Institute of Technology (Eidgenössische Technische Hochschule) in Zürich, made the constructs and performed the homologous recombination and selection of ES cells in Freiburg, whereas for the blastocyst injection Köhler had to cooperate with Horst Bluethmann who had the microinjection techniques established at Hoffmann-LaRoche in Basel. With the production of knockout mice Köhler left the antibody theme and turned to the cytokine genes. Marinus Lamers, group leader in Köhler's department and pursuing the elusive role of IgD, later produced δ chain knockout mice but Köhler was no more than marginally involved in this work. The reasons for his change of interest are not really clear. It is unlikely that he was discouraged by Rajewsky's head start as Köhler's first cytokine knockout mouse, a deletion of the IL-4 gene, was an exact replica of a knockout mouse produced almost 2 years earlier by Werner Müller in Rajewsky's department. Rather, Köhler probably did not care very much what gene to choose and took whatever seemed of current interest. After their initial discovery in the early 1970s as key regulators of the immune system cytokines had a second wind ever since Tim Mossmann and Robert Coffman had first observed that T helper cells fell into two types that could be distinguished by their patterns of cytokine secretion. Some T helper cells secreted IFN-γ, TNF-β and IL-2, and were dubbed Th1 cells, while other T helper cells secreted IL-4, IL-5, and IL-10, and were termed Th2 cells. Mossmann's and Coffman's discovery made a lot of sense since it had been known for many decades that the immune system could respond to antigens in two different ways, earlier referred to as cellular and humoral immune responses. In the latter, antibodies are produced which attach to microorganisms living in body fluids and neutralize them in various different ways. In the former, macrophages and other cells become activated so they can kill bacteria which are inaccessible to antibodies because they have adapted to live inside cells. While many immune responses combine both alternatives, certain infectious agents selectively induce responses which are strongly polarized in one or the other direction. How does the immune system decide which of the two responses was the appropriate one, and how does it keep the two types of responses separate? Now it seemed that the two types of responses corresponded to two types of T helper cells, i.e., the cytokines secreted by Th1 cells were responsible for macrophage activation, whereas the cytokines secreted by Th2 cells stimulated B cells to produce antibodies. In addition, *in vitro* studies on the action of the cytokines involved had revealed a reciprocal negative feedback such that IL-4 inhibited Th1 cells and IFN-γ inhibited Th2 cells. It was thus interesting to study immune responses in knockout mice in which one or the other cytokine gene had been disrupted. The phenotype of the IL-4 knockout mice confirmed and extended current concepts of Th1/Th2 polarity. As expected, there was no production of IL-4 but production of IL-5 and IL-10 was reduced as well, indicating that IL-4 played a master role controlling the level of the other Th-2 cytokines. IFN-γ secretion was enhanced, in line with the feedback inhibition of IL-4 on Th1 cells. With respect to antibody production the results revealed a more complicated situation than expected, i.e., the different Ig classes depended to different degrees on IL-4. Only IgG1 and IgE were strongly reduced in the knockout mouse, whereas other IgG classes and IgA were normal or enhanced. Köhler's paper was published in 1993 in *Nature*, almost 2 years after Müller's first report on IL-4 knockout mice in *Science* in 1991. The fact that *Nature* agreed to publish a paper describing what was practically a repeat experiment is amazing and reflects the sensational quality of the knockout technology at its beginning.

In the controlled microbial environment of the experimental animal facility the lack of IL-4 caused no apparent health defects in the IL-4 knockout mouse. Thus, a meaningful analysis of the role of IL-4 in protective immunity required experimental models of infections with pathogens. Over many years and the work of many laboratories, a number of model infections had been established in mice in which the protective immune response depended on defined components of the immune system, such as Th1 or Th2 cells, T killer cells, antibodies, etc. One of these models is the infection with the nematode *Nippostrongylus brasiliensis*, an extracellular parasite inducing a protective immune response which consists of two components, IgE antibody production and an increase in eosinophilic leukocytes, both dependent on Th2 cytokines. The eosinophils armed with specific IgE antibodies attack the worms and kill them. Upon infection of IL-4 knockout mice with the nematode both components of the immune response were greatly impaired, demonstrating an important role of IL-4 in protection against the worm. Indeed, the results obtained in this model infection uncovered the most impressive phenotypic changes in the IL-4 knockout mice. It must have come as an unpleasant surprise to Köhler that the characterization of cytokine knockout mice required expertise in experimental infectiology, a field of research for which he had held nothing but contempt only a few years earlier in the controversy about the future of the MPIIB (see Chapter 11). Ironically, now that most activities in the area of infection had been terminated in the MPIIB, Köhler's work began to depend on these models. For worm infections Köhler had to turn to experts at Ciba-Geigy in Basel for cooperation. One of a few immunologists in the MPIIB working on infection was Jean Langhorne, who studied a model of mouse malaria. The protective immune response of mice to this parasitic infection consists of two phases, a Th1 phase is followed by a Th2 phase and finally by parasite clearance. In IL-4 knockout mice, the Th2 phase was delayed, resulting in higher parasite levels and delayed parasite clearance than in control mice, but the knockout mice finally also cleared the parasite. The results demonstrated that the importance of a cytokine may vary from one type of infection to another.

One of the most thoroughly studied model infections in the field of Th1/Th2 polarity has been the infection of mice with *Leishmania major*, a tropical parasite transmitted by sandflies that induces cutaneous or intestinal Leishmaniasis in humans. Shielded from antibodies, the parasite lives in macrophages and a protective immune response depends on macrophage activation by cytokines produced by Th1 cells. Today we believe that humans which are able to mount an effective Th1 response develop cutaneous Leishmaniasis, the milder of the two forms. Patients unable to mount a strong Th1 response come down with visceral Leishmaniasis, also termed Kala Azar and often fatal. The reasons why some individuals develop stronger or weaker Th1 responses are not all known but include genetic disposition as an important element. This seems to be reflected in inbred strains of mice, which differ in their sensitivity to *Leishmania major* infection. Infection of mice of strain B6 or strain 129 with *Leishmania major* results in minor skin lesions which heal after a few weeks. In contrast, mice of strain BALB/c develop severe lesions that do not heal and eventually kill the mouse. The difference in sensitivity to *Leishmania* infection between the two strains had been a subject of speculation for many years until the concept of Th1/Th2 polarity provided the basis for a reasonable hypothesis: Among the many available inbred strains of mice there are some with a genetic tendency towards a Th1 response, for example strains B6 or 129, whereas others are genetically predisposed to respond with Th2 cells, for example strain BALB/c. The availability of the IL-4 knockout provided a unique and novel approach to examine this hypothesis. If BALB/c mice were predisposed to a Th2 response, they would respond to infection with *Leishmania* by producing large amonts of IL-4, thus preventing their Th1 cells from secreting IFN-γ which is needed for a protective immune response against the intracellular parasite. BALB/c mice with a deleted IL-4 gene should therefore respond with a greater production of IFN-γ, making them more resistant to *Leishmania* infection than wild-type BALB/c mice. Experiments to test this prediction were done by two groups of investigators, Kopf and Brombacher in Köhler's department and Nancy Noben-Trauth at the National Institute of Health in Bethesda, in collaboration with Ingrid Müller of the Imperial College in London, an experienced infectologist. Both papers were published in 1996, one year after Köhler's death, with seemingly contradictory results. Noben-Trauth had been a guest investigator in Köhler's department to learn the knockout technology and had moved on to the NIH. There she had produced IL-4 knockout BALB/c mice and Müller in London tested their resistance to *Leishmania*. Their results were published under the title: "Susceptibility to *Leishmania major* infection in interleukin-4-deficient mice". A few months later, Kopf and Brombacher published their results under the title: "IL-4-deficient BALB/c mice resist infection with *Leishmania major*". There was a considerable international irritation about this discrepancy. Comparing the actual results of both groups one saw that they were not as clearcut as the titles seemed to suggest. In Kopf's paper the IL-4 knockout BALB/c mice reached only an intermediate level of resistance compared to fully resistant 129 mice. In Noben-Trauth's and Müller's paper, the IL-4 knockout BALB/c mice showed marginally increased resistance compared to wild-type BALB/c mice. Nevertheless, the data were still discrepant and the actual reasons have never been elucidated. Pasquale Kropf, then PhD student in Müller's laboratory who did the experiments, is convinced that Kopf and Brombacher did not use good quality parasites for their infection experiments. In the end, the affair did not destroy the paradigm on the role of Th1/Th2 polarity in the genetic control of the resistance of mice to *Leishmania major*. Müller maintains that the situation is more complicated than that, but the scientific community likes to follow the simplifiers and not the complicators.

Only 1 year after the IL-4 knockout Kopf had completed the knockout of IL-6, a multifunctional cytokine produced by macrophages with suspected roles in hemopoiesis, inflammation, and the immune response. The involvement of IL-6 in many biological processes required the analysis of the knockout mice in a number of different biological responses including bacterial and viral infections. In addition to a reduction in hemopoetic stem cells, the knockout mice showed impaired immunity against many infections, including vesicular stomatitis virus and vaccinia virus, both Th2- and antibody-dependent responses, the bacterium *Listeria monocytogenes*, a Th1-dependent response, whereas the T killer cell-dependent response to lymphocytic choriomeningitis virus was normal. Moreover, the acute phase inflammatory response to tissue damage was severely compromised. Following tissue damage, the liver synthesizes a number of so-called acute phase proteins the common function of which is to counteract the consequences of tissue distruction in various ways. Experimental tissue damage was induced by injection of turpentine into the peritoneal cavity of the mouse. Production of acute phase proteins was strongly diminished in IL-6 knockout mice, demonstrating a role for IL-6 as a messenger between the damaged tissue and the liver. The work on the IL-6 knockout mouse involved several groups, incuding the laboratory of Rolf Zinkernagel in Zürich in which most of the infection experiments were done, and that of

Tadamitsu Kishimoto in Osaka, who in the past had contributed most of the knowledge on the role of IL-6 in inflammation, and who did the analysis of the acute phase response. A conflict developed between Köhler and Kishimoto about the availability of the mice. During the course of the work, Kishimoto had written to Bluethmann, in whose laboratory the mice had been initially produced by blastocyst injection, asking if he could collaborate with Bluethmann in characterizing the mice. Köhler was afraid that he could be bypassed by a shortcut between Kishimoto and Bluethmann and wrote an angry letter to Bluethmann, firmly stating that only he, Köhler, was in control of access to the mice and approval of cooperations. Kishimoto wrote back to Köhler and apologized with Japanese formality.

Köhler did not live to see publication of the paper describing the IL-5 knockout mice which Kopf generated in Freiburg and completed characterizing at the BII where he went as a postdoctoral fellow. The mice had a rather subtle phenotype. Cytokine production including IL-4 and IFN-γ, as well as the distribution of immunoglobulin classes, were unaltered, suggesting a minor role of IL-5 in Th1/Th2 polarity. However, a particular subset of B cells that resides in the peritoneal cavity developed more slowly, and the knockout mice did not respond with an increase in eosinophilic leukocytes to a helminth infection. The latter defect, however, did not impair clearance of the wormload, presumably because the IgE antibody response was normal in the knockout mouse. Another knockout, deleting a cell surface antigen suspected to have a role in cell adhesion, was completed after Köhler's death by his collaborator Peter Nielsen.

# Chapter 15
# Köhler's death

In the morning of 1 March 1995, arriving for work I ran into Rolf Kemler on the parking lot of the MPIIB, who informed me that Georges had died during the night. Kemler, then acting director, had been called by Claudia in the early morning. We all knew that Georges had suffered from heart disease for several weeks, but he had not been hospitalized and had not been on sick leave for any length of time. As a result, the news of his sudden death came as a shock to us all.

Georges had begun to feel ill in November 1994. He noticed that he was short of breath while climbing stairs, could not exercise and rapidly tired out when he walked even on flat surfaces. His private practitioner referred him to Professor Hanjörg Just, chairman of the Cardiology Unit of the Department of Internal Medicine of the Medical School in Freiburg. On first examination, Just noted a pronounced enlargement of the heart, irregular heartbeat, and very low blood pressure. X-ray and ECG were taken as well, the latter showing pronounced pathology but no overt evidence of a coronary event. Just informed Köhler that his condition could be serious and suggested to have him admitted to the hospital for further diagnostic procedures to be performed. Köhler refused hospitalization but agreed to have a 24 h ECG done at home. He came back to see Just a second time in December to discuss the results, and Just again strongly recommended Köhler to be hospitalized for a left heart catheter examination so that his condition could be properly diagnosed. This examination is done by surgical insertion of a tube with a multifunctional device on its tip into the femoral artery under local anesthesia, followed by pushing it against the bloodstream through the aorta all the way upwards to the left atrium and ventricle, under visual control on a screen by computer tomography. The procedure helps diagnose most of the common pathologies of the heart and to do a coronary angiography, as well as therapeutic dilatation of a coronary or insertion of a stent, if necessary. The procedure is not free of risk, in the worst case cardiac arrest can occur. Köhler refused again and never came back to Just, but accepted to take medication that Just prescribed, including digitalis and an acetylcholine esterase (ACE) inhibitor. Just today: "Köhler refused to listen to most of my recommendations, he was almost hostile. He seemed unable to accept the fact that he was seriously ill."

In the following time until his death Köhler was in the care of a private practitioner, a cardiologist and specialist of internal medicine, who had access to the facilities of the Cardiology Center in Bad Krozingen, a recognized private hospital close to Freiburg that specializes in heart transplantation. Upon first examination on 4 January 1995, using various diagnostic procedures, the cardiologist diagnosed a cardiac insufficiency according to the New York Heart Association (NYHA) grade IV, the most serious grade on the scale. All parts of the heart were drastically enlarged, the left atrium and ventricle more than the right. The heartbeat was too fast and irregular and the blood pressure was dangerously low. There was an exudate in the pericardial space, a symptom which often accompanies a myocarditis of infectious, possibly viral, origin. Auscultation revealed noises of the lung consistent with pneumonia and edema. Due to the extended heart muscle the mitralis and tricuspidalis heart valves were unable to close and blood regurgitated at every contraction. Altogether Köhler was in a life-threatening condition. Nevertheless, he still refused to stay in the hospital and went back home against the advice of the cardiologist. During January, his condition was controlled at several time points and seemed to ameliorate. On 7 February the cardiologist judged his heart insufficiency as NYHA grade III. The pericarditis and the pulmonary symptoms had receded, and the blood pressure had somewhat improved. Nevertheless, the diagnosis still was "pronounced diffuse left-ventricular myocardial damage with left-ventricular dilatation, modest right-atrial dilatation, relative mitral regurgitation (grade 1), minor tricuspidal regurgitation, complex ventricular extrasystoly (Lown II to IV)." Köhler then accepted the suggestion of his cardiologist to have a right heart catheter examination performed at the Cardiology Center in Bad Krozingen. In this procedure the device is inserted intravenously and carried with the bloodstream to the right atrium and ventricle. Complications occur almost never but the diagnostic possibilities are limited to the right heart and the pulmonary part of the circulation. The procedure was performed on 20 February and revealed a pronounced and rapid pathological increase in cardiac pressure upon minimal exercise, while the hemodynamic parameters at rest had nearly normalized. This was the last contact that Köhler had with his physicians. They again recommended him to get hospitalized for further diagnostics including a left heart catheter examination with coronary angiography as the only way to diagnose the origin and exact nature of his condition. Again Köhler could not be persuaded. A week later he died in his bed, next to his wife, in the early hours of the morning. A reanimation team arrived shortly but did not succeed.

Georges had not concealed his condition but most of us had not been aware how seriously ill he was. He had tried to lead a relatively normal life by coming to the institute nearly every day and for the full day. On 13 February Rolf Kemler had invited everybody to a party celebrating his 50th birthday. The Köhlers attended and George danced with my wife, Barbara, one of the slow dances. He told her how unhappy he was about his illness, and that he had to take it easy with the dancing. He also said that he was on his way to improvement and hoped to reach a full recovery. He may indeed have felt in better condition, because he had agreed to attend a meeting of the scientific advisory board of the German Museum which was planned for 3 March in Munich, asking Dr. Frieß for hotel reservations to be made for himself and Claudia. While this may be evidence for an optimistic attitude, Melchers recalls that Georges visited him in Basel during the last week of February to bring some mice, an errand that he normally would have delegated to a co-worker. During the visit Georges went to say hello to many of his colleagues in the BII, a behavior not typical for Georges. Melchers today: "Georges may have felt his death coming. He may have wanted to say goodbye."

The cause of Köhler's death was subject to wild speculations in the public domain. Numerous times I received phone calls from journalists asking if he had died of AIDS. Another rumor that was frequently reported was that Köhler had refused to accept a heart transplant. That was clearly not the case, because his physicians had not been in the position to suggest heart transplantation for lack of precise diagnostic information. Fact is that Georges died of sudden heart failure, but what caused his heart disease remains unclear. According to Just, the symptoms of pulmonary infiltration together with the pericardial exudate point to a myocarditis of viral origin, but a coronary insufficiency could have caused similar symptoms and thus cannot be excluded.

What made Georges resist hospitalization and proper diagnosis remains subject to speculation as well. I do not think that he was afraid of invasive diagnostic procedures, as I do not remember Georges as a timid person. Two non-mutually exclusive possibilities remain as explanations in my opinion. On the one hand, he certainly lacked the knowledge and may also have lacked the imagination to judge the graveness of his disease. Together with his notorious disregard for opinions and advice by others this may have led to a deadly mixture of obstinacy and carelessness. On the other hand, Georges may have had a clear imagination how his life would develop if he recovered from his illness as an invalid with chronic cardiac insufficiency or as a recipient of a heart transplant. His reaction may have been to minimize the likelihood of such an outcome by purposely putting his life at risk.

While Claudia wanted the funeral to take place with only the family present, she and the three children attended the memorial ceremony for Georges at the MPIIB on 13 March. Until today, Claudia has kept a loose relation to the institute, occasionally attending social events such as Christmas parties, the celebration of the 40th anniversary of the MPIIB, and the dinners on the occasions of the Georges Köhler Lectures that are held annually at the MPIIB by invited prominent immunologists. Their youngest son, Fabian, has recently obtained a PhD degree of the University of Freiburg with a thesis done at the MPIIB under the supervision of Michael Reth, and is now working there as a postdoc. For the most part, Claudia's life continues to be family oriented, concentrating on two grandchildren from her daughter Lucie who lives in Freiburg. When the city administration asked her if she would agree to having the novel Freiburg Biotech Park named after her husband, she refused. However, she agreed to a street in Freiburg being named "Georges Köhler Allee".

There was extensive coverage of Köhler's death in all major German newspapers, again revealing his star-like quality reaching out beyond the scientific domain. The German Museum in Bonn dedicated their first catalogue to Köhler, who died before the Museum was officially opened in 1996. The German Society of Immunology established a prize in his name, for outstanding contributions by scientists at the postdoctoral level. The MPIIB established a new junior research group and named it after him. Multiple times I wondered why Claudia so vehemently refused to share some of the impressions and recollections of her life with Georges with me for inclusion into this book. A few weeks after his death she had asked Barbara the rethorical questions: "Was it not a clever move of him to die at this point in his life? Was it not perfect timing?", obviously meaning that after having reached the summit of his career so early the path could only have led downwards. I tend to agree with Melchers who surmised that Claudia wishes to maintain Georges's memory as a celebrity undisturbed, that she may want to avoid having to reveal her candid opinion about Georges as a human being with rather normal talents and abilities. Melchers thinks that Georges's career may have been rather ordinary had he not invented monoclonal antibodies: "Chances are that he would have blended in with the majority of unknown research scientists, doing unspectacular work in some university laboratory."

## Chapter 16

# Magic bullet

The first demonstration of the amazing advantage of monoclonal antibodies in a biotechnology procedure came from Milstein's laboratory. By the late 1970s essentially everybody in Milsteins laboratory had become involved, in one way or another, with making and studying monoclonal antibodies. David Secher, postdoc in the laboratory, tested the use of monoclonal antibodies in affinity chromatography. In this technique, introduced earlier by the Nobel Laureate biochemist Christian Anfinsen to purify substances, antibodies specific for the substance to be purified are covalently coupled to microbeads and the material is then packed into a chromatography column. A fluid that contains a heterogeneous mixture of molecules including the substance in question is then slowly filtered through the microbead column. Upon passage through the column, the targeted substance – but ideally no other compound in the fluid – attaches to the antibodies on the microbeads, and the effluent is depleted of the substance. To recover the substance from the column, one can use several different methods, for example acid pH or high salt, that denature the antibodies so that the attached substance gets released and can be washed out of the column with appropriate solvent. Often the denaturing conditions can be chosen to be reversible so that the column can be renatured and used again. Using conventional antibodies which contain a mixture of immunoglobulins with all sorts of binding specificities in addition to the desired one, the targeted substance can be more or less enriched but there are always impurities which are copurified by binding to these unknown antibodies. Secher concentrated on purifying α-interferon, a molecule that had originally been discovered by the Swiss virologists A. Isaacs and J. Lindenmann and held strong promise for potential use in human medicine. They had observed that when certain cells were infected with viruses, a substance was released that conferred an antiviral state when added to new cultures of the same cells. After pretreatment with the supernatant of infected cultures, new cultures could resist infection with the virus. Of course, the nature of the substance was of prominent interest, but its molecular characterization required purification of the material, which turned out to be difficult with conventional procedures, at best leading to preparations with enhanced activity. Such preparations still contained a lot of different molecules so that it could not be determined which of them had the antiviral activity. Secher immunized mice with such impure preparations and produced a series of monoclonal antibodies, each reacting with one of the various components present in the preparation. By testing the monoclonal antibodies one by one for their ability to neutralize the antiviral activity, Secher identified an antibody reactive with the substance of interest. He then coupled the antibody to microbeads and performed affinity chromatography by passing culture supernatants with antiviral activity over his columns. To Secher's delight, elution yielded a single pure protein by which the antiviral state could be transferred – α-interferon. Secher today: "We patented the procedure, again against the advice of the MRC patenting authorities. The patent has never been successfully challenged until today." While α-interferon today is an important component in the treatment of human viral infections such as hepatitis B and C, it is no longer produced by Secher's method, useful as this has been in the identification process. Recombinant gene technology and expression in bacteria or mammalian cell lines is now the method of choice in the production of biomolecules for medical use.

The biotechnology industry today rests to a large extent on three pillars, the production of biomolecules in bacteria or mammalian cells by recombinant gene technology, the production of recombinant plants for agriculture, and monoclonal antibodies. Monoclonal antibodies are widely used as reagents in laboratory research and as diagnostic reagents in laboratory medicine. Monoclonal antibodies are now commercially available to detect essentially any known molecule present within cells or on their surface, and cutting edge research in cell biology, biochemistry, and immunology is unimaginable without these tools. Similarly, monoclonal antibodies to blood constituents, hormones, cytokines, or other mediators are indispensa-

---

Figure 28
Purification of α-interferon by affinity chromatography using a monoclonal antibody. A tissue culture supernatant enriched for interferon is injected into mice to induce antibodies. The supernatant contains a large number of different substances but very little interferon, as seen by electrophoretic analyses A/B. The electrophoresis A reveales only the major protein of tissue culture fluid, albumin, as a single high-molecular weight band, but no band for interferon. Electrophoresis B was overloaded and reveals multiple unseparated components with a faint band in the position of interferon (arrow). Spleen cells of the immunized mouse were fused with myeloma cells and the cloned hybridomas tested for antibodies that neutralize interferon activity. A positive hybridoma was injected into mice to produce large amounts of the monoclonal antibody which was purified from the serum, coupled to beads, and the beads packed on a column. The supernatant was passed through the column, the unbound materials washed through, and the retained material subsequently eluted by mildly acid pH. Electrophoresis of the eluate showed three bands (C), the strongest band corresponding to interferon. After repeating the procedure the eluted material (D) consisted of pure interferon (from Milstein, 1980).

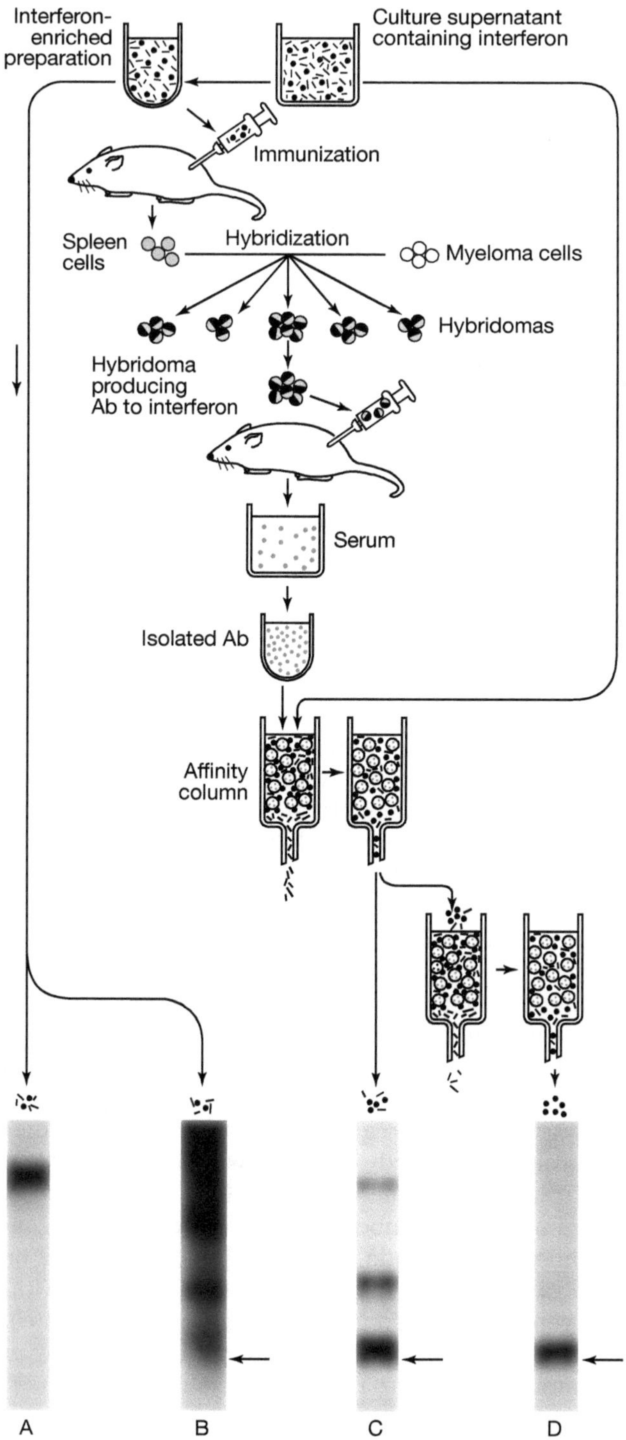
Interferon-
enriched
preparation
Culture supernatant
containing interferon
Immunization
Spleen
cells
Hybridization
Myeloma cells
Hybridomas
Hybridoma
producing
Ab to interferon
Serum
Isolated Ab
Affinity
column
A
B
C
D

ble today in clinical diagnosis. This sector has rapidly grown with little delay after the hybridoma technology became available, and the applications are so numerous that a full account would fill several volumes of the size of this book. Several examples have been discussed in Chapter 8. Users often do not know that the products they purchase are based on monoclonal antibodies. For example, the pregnancy tests that women buy in a pharmacy to test their urine are based on monoclonal antibodies specific for the human pregnancy hormone, chorion gonadotropin, and assembled with reagents to give a color reaction upon binding the hormone. The majority of blood tests that medical laboratories do for diagnostic purposes are now based on monoclonal antibodies. The worldwide sales of monoclonal antibodies was around 3.5 billion US$ in 2001, and grew to an estimated 5 billion US$ in 2003, and is expected to at least double by 2008.

Around the time of the Nobel Prize for Jerne, Köhler, and Milstein the term "magic bullet" has become a synonym for the expectations of the public in monoclonal antibodies as weapons in the battle against cancer. Although the use of monoclonal antibodies in the therapy of human diseases has taken considerable time to develop, a number of therapeutics based on monoclonal antibodies are now on the market after yielding promising results in clinical studies. One reason for the delay was that antibodies produced in mice, when injected into humans, elicit immune responses because the human immune system recognizes mouse antibodies as foreign antigens. The reaction is known from the ancient times of serum therapy (see Chapter 1) when patients suffering from tetanus or diphtheria developed fatal "serum sickness" when they were treated repeatedly with antibodies from the same animal, for example horse. The patients developed antibodies to horse immunoglobulins after the first treatment that would bind to the horse antibodies upon subsequent injection, causing anaphylactic shock and sometimes death. Severe serum sickness could be avoided by using antibodies from another animal, for example swine, upon second treatment. The antibodies to horse immunoglobulins developed by the patient after the first treatment would not or only weakly react with the swine antibody used for the second treatment. A similar strategy was, however, not feasible for treatments of patients with monoclonal antibodies, as the hybridoma technique was established for mice and rats, and target diseases such as cancer likely required chronic injection of the same monoclonal over prolonged periods of time. Two strategies were developed by researchers to circumvent the dilemma, the production of chimeric antibodies and of humanized antibodies, both by recombinant gene technology. Chimeric antibodies are made by recombining the mouse or rat variable region genes isolated from the appropriate hybridoma with human constant region genes. This is done for both heavy and light chains, and the recombinant genes are introduced into a cell line that now produces antibodies with mouse or rat variable regions, contributing the specificity for the antigen, on human constant regions. The advantage of the human constant regions is two-fold. They are tolerated by the human immune system, i.e., they do not elicit immune responses, and they exert the effector functions in the human host even better than mouse constant regions, such as reacting with complement or with surface immunoglobulin receptors on cells able to kill tumor cells. However, chimeric antibodies still can induce immune reactions in the patient. The variable regions are of non-human origin, and both the framework and hypervariable sequences are recognized as foreign. A further improvement can therefore be achieved by humanization of a mouse or rat monoclonal antibody. In this approach, only the hypervariable regions are derived from the hybridoma, while the framework sequences in the variable regions as well as the entire constant regions are of human origin. While there is still the possibility that the patient produces antibodies to the hypervariable regions, so-called anti-idiotypic antibodies (see Chapters 1, 3 and 4), the problem appears to be bearable and both chimeric and humanized antibodies are presently licensed for human use. Another delaying element in the therapeutic use of monoclonal antibodies, particularly against cancer, has been that investigators could rarely, if at all, identify candidate antigens on cancer cells that are not also expressed on at least some normal cells in the body. Monoclonal antibodies against cancer cells would thus always target normal cells as well and possibly harm them. While this is clearly the case, experience has told that these problems are manageable and rarely worse than the adverse side effects of conventional cancer therapies such as chemotherapy or irradiation. In 2002, the market of monoclonal antibody-based therapeutics amounted to about 16% of the global biotechnology market, with a strong expectation to expand, as a host of new monoclonals are in the process of development and clinical evaluation. Main target disorders for monoclonal antibody therapies are several forms of cancer, including their ability to induce the formation of blood vessels, chronic inflammatory diseases, and adverse reactions of the immune system such as allergies and transplant rejection.

The first monoclonal antibody for treatment of a human cancer was approved by the FDA in 1997. It is manufactured by Genentech and distributed under the trade name Rituxan® (Rituximab). The antibody is chimeric, i.e., it consists of mouse variable and human constant regions, and is used to treat non-Hodgkin's B cell lymphomas. Non-Hodgkin's lymphomas are a diverse group of B cell malignancies which collectively rank fifth in cancer incidence and mortality. Rituxan® binds to a surface antigen, CD20, which is found on most B cells including normal B cells and B lymphoma cells. Its exact mode of action is unknown, but it is likely to induce so-called "antibody-dependent cellular cytotoxicity" (ADCC) against the cancer cells. ADCC is a mechanism which the immune system uses as defense against bulky aggressors such as parasites and worms, but also cancer cells. Antibodies, which by themselves cannot harm large organisms, bind with their constant region to immunoglobulin receptors on the surface of specialized leukocytes and macrophages which have the ability to kill target cells. The variable region of the antibody sticks out and is free to attach to a parasite or cancer cell, depending on its specificity. The attachment brings the cytotoxic leukocyte in contact with the target cell or organism so that killing can occur. Additional modes of action include the activation of the complement system which can also lead to destruction of target cells following binding of antibody. Because expression of CD20 is not restricted to B lymphoma cells Rituxan® is expected to attack normal B cells as well, causing side effects of immunodeficiency. However, Rituxan® is often given in combination with chemotherapy and does not increase the immunodeficiency caused by chemotherapy alone. Conventional treatments of non-Hodgkin's lymphoma vary with the state of progression upon first diagnosis. As long as the disease is asymptomatic physicians usually wait and do nothing. Low grade disease is treated by either single agent oral chemotherapy or a combination of cyclophosphamide, vincristine, and prednisone (CVP). Intermediate grade non-Hodgkin's lymphoma requires high intensity chemotherapy by including, in addition to CVP, anthracycline and doxorubicine (CHOP). Rituxan® is efficient in patients with low and intermediate grade disease. In a single agent multicenter study on 151 low grade patients, about 50% responded to Rituxan® therapy with partial (44%) or complete (6%) remission and an average delay of disease progression of 13 months. Given to intermediate grade patients in combination with CHOP, Rituxan® increased the rate of complete remissions from 60% to 75%, 69% of patients remained free of relapse compared to 49%, and the overall response rate increased from 68% to 83%, compared to patients receiving CHOP alone. Rituxan® was the first drug ever that could be shown to improve the therapeutic effects of CHOP in the treatment of non-Hodgkin's lymphoma. The demonstration of its beneficial effects was a tremendous boost for the development of further monoclonals for cancer therapy.

Soon after the success of Rituxan® additional anticancer therapies based on monoclonal antibodies were approved by the FDA and appeared on the market. Oncolym® (Lym-1) is manufactured by Peregrine Pharmaceuticals and binds to the human MHC class II molecule HLA-DR, which is expressed on B cells but also on other cells in the body, including dendritic cells which are of pivotal importance in the initiation of immune responses. The monoclonal antibody is unmodified, i.e. of complete mouse origin, and is coupled to the radioisotope $^{131}I$ for clinical use. By the binding of the antibody to target cells, a high concentration of $^{131}I$ accumulates in their vicinity, causing damage of the DNA in the target cells. Like Rituxan®, Oncolym® is in use for the treatment of non-Hodgkin's B cell lymphomas that often express high levels of HLA-DR. Clinical tests resulted in about 50% partial and 7% complete remissions in low grade and 13% partial and 5% complete remissions in intermediate grade non-Hodgkin's lymphoma. Oncolym® is an example of the use of a monoclonal antibody to deliver a toxic agent to the target cell, in contrast to Rituxan® whose action is based on the natural effector functions of the antibody alone. Both therapeutic principles appear to be feasible.

Wyeth-Ayerst Laboratories developed Mylotarg®, a combination of a monoclonal antibody (Gemtuzumab) and a chemotherapeutic agent (Ozogamycin), for the treatment of acute myeloid leukaemia, a malignancy of the myeloid lineage affecting mostly older people. The monoclonal antibody is humanized, i.e., the hypervariable regions of the mouse heavy and light chains are grafted into a human antibody of the IgG4 class, and is produced in mammalian cell cultures. It binds to the surface molecule CD33, an adhesion molecule expressed on myeloid leukemia cells and normal progenitor cells of the myeloid lineage, but not on hemopoietic stem cells. The antibody is covalently coupled to the bacterial compound chaliceamycine which kills exposed cells by inducing double-stranded DNA breaks. Mylotarg® is thought to act by destroying the leukemia cells by concentrating chaliceamycine on their surface through the binding of the antibody to CD33. While normal myeloid precursor cells may be destroyed as well, the important hemopoietic stem cells, responsible for replenishing all other blood components, are not affected. Mylotarg® has first been tested on patients who suffered from a relapse of the disease after chemotherapy. Of 104 such patients, about one out of three responded with a remission, a rate comparable with that of standard chemotherapy regimens, but with significantly fewer adverse side effects. While the company initially advised against using Mylotarg® in combination with chemotherapy, a more recent study tested the effect of a 21-day regime combining Mylotarg® with the chemotherapeutic agent Genasense® in the treatment of older patients with relapsed acute myeloid leukemia. Complete remissions were seen in 12 out of 39 patients with 10 surviving more than 6 months. Mylotarg® is another example of the use of monoclonal antibodies as vehicles to deliver a toxic agent to the target cell, also referred to as "immunotoxins".

Campath® (Alemtuzumab), manufactured by ILEX Pharmaceuticals and distributed as MabCampath® in Europe, binds to CD52, an antigen expressed on several types of white blood cells including B cells. It is a humanized monoclonal antibody in which the hypervariable regions stem from a hybridoma originally produced in a rat. It is so far the only monoclonal antibody licensed for the treatment of relapsed chronic lymphocytic leukemia (CLL), a B cell malignancy, but is also considered and tested in clinical trials for other types of leukemia. CLL is the most frequent form of leukemia in adults, affecting about 120,000 people in Europe and the US. Campath® was tested in three clinical studies on a total of 149 CLL patients who had failed to respond to fludarabine, the standard chemotherapeutic agent for CLL. Partial responses were between 21–31%, complete responses were 0–2%, with mean relapse-free survival times of 4 to 7 months.

Monoclonal antibodies have been developed for treatment of solid tumors as well. Perhaps the most widely used is Herceptin® (Trastuzumab), distributed by Roche AG and approved by the FDA in 1998, which is effective against breast cancers that overexpress a certain growth factor receptor, termed HER2/neu or erb-B2. Erb-B2 is one of four members (erb-B1-4) of a family of receptors for the epidermal growth factor (EGF), a cytokine that binds to its receptors and regulates cell division of epithelial cells including that of the mammary gland. The mammary gland is a highly dynamic tissue which is subject to dramatic functional changes during the female hormonal cycle, pregnancy, and lactation. In this context, epithelial cells are turned over by cell death and cell division, a process that needs to be regulated to remain in balance. In addition to the EGF receptors, hormonal receptors have a role in this complex process, most prominently estrogen receptors. Dynamic tissues with a high turnover have a higher rate of malignancies than tissues with slow turnover, because the frequent cell divisions provide opportunities for the cells to accumulate mutations, some of which predispose the cell to become cancer cells. One of these mutations is a selective multiplication of the gene for erb-B2 resulting in overexpression of the EGF receptor and accelerated growth. About one-third of breast cancers show this type of mutation. Herceptin® is a humanized monoclonal antibody that binds to erb-B2 and blocks its accessibility to EGF. As a result, tumor growth may be slowed down. In addition, Herceptin® may also activate ADCC like other monoclonal antibodies. Before a breast cancer is treated with Herceptin® a biopsy needs to be taken to determine the expression of erb-B2 on the tumor cells by immunocytochemistry or the degree of amplification of the erb-B2 gene by a hybridization technique termed FISH. Only cases with strong overexpression or amplification can be expected to respond to the treatment. Clinical studies on such selected patients with the chemoterapeutic agents Paclitaxel® or Docetaxel® with or without Herceptin® showed two-fold or more increased overall response rates of the combination compared to chemotherapy alone. Moreover, Herceptin® increased the duration of efficacy from about 4 months to 8 months, and the time to tumor progression from 3–6 months to 7–10 months. In a study with Herceptin® alone given to women who had relapsed tumors after chemotherapy, the response rate was 14%, the duration of efficacy was 9 months, and the increase in survival time was 13 months.

Erbitux® (Centuximab) is based on a similar principle as Herceptin®. This chimeric monoclonal is distributed by Bristol-Myers Sqibb, and binds to erb-B1, another member of the EGF receptor familiy. Erbitux® was approved by the FDA in 2004 for the treatment of colon and rectum carcinoma, the third most common cancer in the US and ranking second in cancer-related deaths. Although no prolongation of survival times were observed in clinical studies involving more than 300 patients, Erbitux® was shown to shrink tumors in about 10% of cases when given alone and in about 20% of cases when given in combination with chemotherapy. Similar to Herceptin®, colon cancers need to be tested for erb-B1 expression before using Erbitux®.

In order to survive and grow, tumor cells need oxygen and nutrients like any other cell in the body. Solid tumors have mechanisms to induce the growth of blood vessels that sprout into the tumor from existing arteries and supply them with blood that transports the oxygen and nutrients into the tumor tissue. The mediators that let blood vessels grow are called vascular endothelial growth factors (VEGFs), a family of cytokines secreted by many tumors. VEGFs bind to VEGF receptors on endothelial cells, a cell type lining the inner surfaces of blood vessels, and make them divide so that new extensions can be formed. Interfering with the ability of tumors to induce their own blood supply is likely to inhibit tumor growth, and so intensive research is going on worldwide to learn how this could be achieved in the most effective way. Among other approaches, a monoclonal antibody against VEGF has been shown to inhibit tumor-induced blood vessel formation and has recently been approved by the FDA for treatment of colorectal cancers in combination with chemotherapy. Termed Avastin® (Bevacizumab), it is a humanized antibody developed by Genentech and distributed by Roche. By binding to VEGF, Avastin® blocks the binding of VEGF to its receptor, thus inhibiting induction of new blood vessels by the tumor. In clinical studies including 815 patients with colorectal cancer Avastin® increased the response rate from 35% to 45%, the time to progression from 6 to 10 months, and the overall survival time from about 15 to 20 months, compared to chemotherapy alone. As induction of new blood vessels is a property of many solid tumors, Avastin® may be effective against tumors other than colorectal cancer as well. Clinical studies to test its effect on other tumors, including prostate and ovarian cancer, are underway.

In addition to tumors, chronic inflammatory disorders respond well to monoclonal antibody therapies, notably rheumatoid arthritis and Morbus Crohn. Rheumatoid arthritis is a chronic inflammation of the joints which affects several percent of the population in industrialized countries, often resulting in severe disability and permanent crippling. In addition to the devastating health impairment of the affected individual, the disease has important socioeconomical consequences. While the actual cause of the disease is not known, involvement of autoimmune reactions appear to play a role, together with genetic disposition and possibly retroviral infection. Conventional treatments are based on anti-inflammatory drugs such as corticosteroids or methotrexate, which may temporarily ameliorate the symptoms but cannot cure the disease. Monoclonal antibodies of several specificities have been considered for therapy of rheumatoid arthritis and tested in clinical trials with variable success, including antibodies to CD3 or CD4, antigens on T cells which appear to be hyperactivated in the disease. The monoclonal antibody that came out as winner in the contest was directed against the cytokine tumor necrosis factor (TNF), a misnomer originating from the way it was originally discovered (see Chapter 9). TNF occurs in two forms, α and β, which are multifunctional mediators that influence innate and specific immune mechanisms in a complex fashion, from the regulation of the development of lymphoid organs in the embryo to the activities of several types of leukocytes during inflammation, tissue damage, and infections. The pivotal role that TNF has in the pathogenesis of rheumatoid arthritis was not fully appreciated until the beneficial effects of a monoclonal antibody to TNF, now marketed under the trade name Remicade® (Infliximab) were discovered. Remicade®, a chimeric monoclonal antibody manufactured by Centocor, was tested in 428 patients with active rheumatoid arthritis that responded poorly to methotrexate. The patients received methotrexate either with a placebo or with Remicade® at two different doses. Their symptoms were graded according to the American College of Rheumatology (ACR) criteria and revealed impressive improvements according to clinical as well as laboratory parameters. Compared to the group receiving placebo, patients achieving 20%, 50%, or 70% remissions were about two-fold, four-fold, and more than 10-fold, respectively, more frequent in the group receiving Remicade®. Improvements were long lasting and 10% of patients receiving Remicade® had completely asymptomatic intervals of more than 6 months, compared to 0% in the placebo group. Remicade® is also effective against Crohn's disease, a chronic inflammation of the large intestines of unknown cause. Although Crohn's disease is not as frequent as rheumatoid arthritis, it is a devastating impairment to the individual patient associated with frequent episodes of diarrhea, abdominal pain and blood loss in the stools. A particularly severe form is fistulating Crohn's disease, in which open connections form between the colon and the outer skin or the rear vaginal mucosa, leaking intestinal content. Conventional treatments of Crohn's disease involve corticosteroids, anti-inflammatory drugs and antibiotics, which may ameliorate the symptoms but do not cure the disease. When given in addition to conventional therapies, Remicade® led to clinical remission in 39–45% of patients compared to 25% in the placebo group. Closure of all fistulae was seen in more than 50% of patients receiving Remicade®, compared to 13% in the placebo group. Remicade® is also effective in several skin diseases including psoriasis. Along with Retuxan®, Remicade® was among the first monclonal antibodies approved for human therapy and its success served a pathbreaking function in the development of further monoclonal antibody therapies.

All therapies with monoclonal antibodies have side effects which are in the majority of cases tolerable when compared to the therapeutic advantages they can cause. However, Remicade® unexpectedly had a dangerous side effect that led to severe restrictions in its recommended use. By mid-2001, about 130 cases of active tuberculosis had been reported worldwide under Remicade® therapy, including several lethal cases. Apparently, neutralization of TNF in the body fluids can cause reactivation of a latent, inactive tuberculosis. Moreover, so-called opportunistic infections have been observed under Remicade® therapy, infections with microorganisms that do not cause disease in immunocompetent individuals. These cases indicate that Remicade® may lead to significant immunosuppression. As of 2002, patients need to be diagnosed for tuberculosis, active or latent, and other ongoing infections before Remicade® can be administered. In addition, Remicade® is not recommended for patients with cardiac insufficiency.

Xolair® (Omalizumab) is a humanized monoclonal antibody developed between Tanox, Genentech, and Novartis, and recently approved for the treatment of allergic asthma. It recognizes immunoglobulins of the IgE class which are important mediators of many sorts of allergies. IgE is a rare antibody class whose beneficial function is to defend against parasites and worms (see Chapter 14). IgE binds to IgE receptors on the surface of certain white blood cells, so-called eosinophilic and basophilic leucocytes that can kill these organisms, and mast cells which release mediators, such as histamine and serotonine, that make capillaries leaky so that white blood cells can leave the blood stream and reach parasites in the tissues. Allergic individuals tend to produce too much IgE in immune reactions, leading to allergies to a variety of different environmental antigens, so-called allergens. If an allergic person is exposed to the allergen, the IgE activates the leukocytes to release their mediators, followed by the well know allergic symptoms, such as swelling and seeping of fluids from mucosal surfaces. In the case of asthma, swelling of the bronchial mucosa leads to bronchial constriction. Xolair® prevents the binding of IgE to the IgE receptors and can thus dampen the symptoms of asthma. While the beneficial effects of Xolair® have been proven in clinical studies, only time will tell if monoclonal antibody therapy will take a firm place in the treatment of asthma and other allergies.

Two monoclonal antibodies have been approved for human use as immunosuppressants to counteract transplant rejection. Xenapax® (Daclizumab) is a humanized monoclonal antibody manufactured by Roche, and Simulect® (Basiliximab) is a chimeric monoclonal antibody produced by Novartis. Both are directed to the T cell surface antigen CD25, one of the components of the receptor for the cytokine IL-2 which T cells need to become activated and to multiply. By binding to CD25, the monoclonal antibodies block the binding of IL-2 to its receptor, thus preventing activation of the T cells involved in graft rejection. Clinical studies in renal transplant patients produced similar results for both antibodies, with a reduction of acute rejections by about one-third compared to standard therapy, but no significant prolongation of long-term graft survival. Again, time will tell if monoclonal antibodies will have a lasting role in transplantation medicine.

This chapter described only the monoclonal antibodies that have so far been approved for human indications by the FDA or equivalent national agencies. Pharmaceutical industries foresee the revenues of the monoclonal antibody market to increase exponentially, and so a much larger number of monoclonals are in their pipelines as therapeutic agents for human use in the future. If one wants to generalize, the experience with the approved applications seems to indicate that a monoclonal antibody can be of significant benefit for 20–50% of patients who receive the monoclonal in addition to conventional therapies, or fail to respond optimally, or relapse after conventional therapies. Thus, while the public expectations expressed by the term "magic bullet" had certainly been unrealistic, a tremendous number of suffering humans can benefit from monoclonal antibody therapies already today, considering the enormous number of individuals who suffer from diseases such as rheumatoid arthritis or cancers. Monoclonal antibodies may not be magic, but bullets nonetheless.

## Chapter 17

# The antibody problem today – not quite solved

Owing to the studies of Tonegawa, Milstein, Rajewsky, and others we know today that B cells, in order to generate antibody molecules of optimal fit to the antigens they encounter, possess a mechanism of somatic hypermutation of antibody genes. At the beginning of an immune response the B cells produce antibodies of the IgM class which mostly carry V regions encoded by unaltered genomic VDJ genes. The diversity in these V regions is determined by the number of the V genes present in the genome but augmented by the V(D)J recombination process. The diversity contributed by V(D)J recombination is considerable, as the joining ends of the gene segments are shortened in a variable fashion during the process, and nucleotides can be inserted randomly by the enzyme terminal desoxynucleotide transferase. However, V(D)J joining diversity is restricted to the third hypervariable regions of heavy and light chains, whereas that of the first and second hypervariable regions is limited by the diversity of the genomic V genes. B cells expressing such IgM receptors are sufficiently diverse so that incoming antigens will always encounter a few B cells with a fitting receptor, but the binding affinity of the secreted IgM is usually low and insufficient for an effective antibody response. Upon repeated exposure to the same antigen the B cells therefore switch to production of IgG receptors which mostly carry V regions encoded by mutated V(D)J genes. The mutations take place while the B cells reside in the germinal centers of the lymphoid organs, the anatomical site where B cells meet with the antigen, get activated and divide, and turn into antibody secreting cells. B cells with mutations that increase the affinity of their antigen binding receptors are activated more efficiently and divide more often than non-mutated B cells or B cells with mutations that lower their binding affinity. As a result, the average affinity of the B cell population increases and high affinity IgG antibodies are produced.

While these results have ended the controversy between germline and somatic mutation theories, they opened another host of questions to be solved. The most pressing of them was how B cells manage to mutate their genes with a thousand-fold higher frequency than that of random mutations. Random mutations mostly occur during DNA replication as copy mistakes. DNA replication is done by a group of enzymes termed DNA polymerases, which rather faithfully match the bases adenin (A) with thymin (T) and guanin (G) with cytosin (C), and *vice versa*, during the replication of both strands of the double stranded DNA. This guarantees that the two resulting double stranded DNAs are identical in sequence to the one serving as matrix. Mistakes are made very rarely, leading to the infrequent random point mutations that occasionally occur. To explain the hypermutation of antibody genes, investigators suspected for a long time that B cells may possess special types of DNA polymerases which make mistakes in base pairing at a high rate. Indeed, it is known from bacteriophages that low fidelity DNA polymerases can be responsible for high mutation rates. As a result, immunologists searched for error prone DNA polymerases in B cells.

However, the clue to somatic hypermutation came from an entirely unexpected angle, as it is so often the case in science. Because the human genome contains a disturbingly small number of genes, the attention of many biologists has turned to diversification mechanisms of RNA to explain what seems to be a discrepancy between the very large number of different proteins in the body and the relatively small number of genes in the DNA. The hypothesis is that RNA molecules are "edited" in different ways so that a single original RNA molecule can serve to code for several proteins. In the context of such studies an enzyme was discovered, termed cytidine deaminase, which turned out to be the key to somatic hypermutation in B cells. Tatsuko Honjo, immunologist in Kyoto, Japan, was among the first to demonstrate a role of cytidine deaminase in hypermutation of antibody V genes as well as in class switching of the heavy chain genes. As mechanism for its action, Honjo proposed that by RNA editing cytidine deaminase edits yet unknown mRNAs to generate specific endonucleases for class switch and hypermutation. In contrast, Michael Neuberger, immunologist in Cambridge, UK, suggested a direct action of the enzyme on the DNA itself: Cytidine deaminase cleaves an amino group off the Cs in the DNA strand, turning them into uracils. Uracil (U) does not normally occur in DNA, rather it is one of the four bases that make up RNA. If a C is replaced by a U, the next round of DNA replication can go one of two ways, depending on what happens first, DNA replication or mismatch repair. In the first case, DNA polymerase replicates over the G-U pair. Since DNA polymerase mistakes the U for a T, the result will be an A-T pair in one and a G–C pair in the other new DNA double strand. In other words, one of the two daughter cells will have a point mutation in this position. In the second case, the U is recognized by enzymes that are specialized to detect mistakes in the DNA and cut them out. As a result, there will be no base in this position, a so-called abasic site is created. Because an abasic site cannot instruct which of the four bases is to be inserted in the complementary position, DNA polymerase will insert randomly any of the four bases in the next round of replication. This process, termed base excision repair, will thus generate a new DNA double strand that contains in that position any of the four basepairs (A–T, T–A, C–G, G–C), in addition to a double strand with the original G-C pair. Again one of the two daughter cells inherits a mutation. To prove the capability of the enzyme to mutate DNA Neuberger expressed it in *Escherichia coli* and found exeedingly high mutation rates in the recombinant bacteria. All the mutations occurred at G–C pairs, as predicted. Furthermore, it was found that cytidine deaminase is expressed at high level in activated B cells that reside in germinal centers, and nowhere else. Hence, cytidine deaminase is expressed in B cells during hypermutation but not in B cells before or after that phase. Together, the observations proved the pivotal role of cytidine deaminase in somatic hypermutation of antibody genes. It was given a new name, activation induced deaminase (AID).

While AID seems to be primarily responsible for mutations at G-C pairs, antibody genes can also mutate at A-T pairs. AID may thus not be the sole hypermutator in B cells. In an attempt to explain A-T mutations as an indirect consequence of AID, Neuberger has proposed a scheme in which the excision of the U can serve as the origin of a larger DNA lesion caused by exonuclease-1 which removes several nucleotides in downstream direction. This gap is recognized by lesion recognition molecules and filled by an error-prone DNA polymerase, such as DNA polymerase-$\eta$ which is known to make frequent mistakes when replicating over A-T pairs. So far, all proposed mechanisms of action of AID are hypothetical, however, and need to be put to experimental examination. An equally fascinating open question in antibody gene hypermutation is what directs the mutations to the antibody V genes, while all other genes in the B cell are protected against hypermutation. Although DNA demethylation and transcription are suspected to have a role in focussing somatic hypermutation to antibody genes, other demethylated and transcribed genes in B cells do not mutate. Imagine the consequences if all the active genes of the B cell would hypermutate. The antibody problem thus continues to keep immunologists running, although the goal seems now in sight.

# Appendix

# Appendix A
# Two lectures given by Georges Köhler to general audiences

## Celebration in honor of Köhler's Nobel Prize by the City of Basel

23 January 1985

Ladies and gentlemen,
in this short talk I intend to tell you about the history of the invention of the hybridoma technique. Some of you may know the technique professionally, many may have read summaries about it in the public press and a few of you may hear about the production of monoclonal antibodies for the first time today. (The others will be called to order by the Capricietto for four timbals right after the talk.) I will try to work out a few points.

The invention of the hybridoma technique is the perfect example that basic research on occasion may have a strong impact on applied research.

The intellectual background that led to the invention of monoclonal antibodies in my view – Cesar Milstein, the co-inventor, may have to tell another story about the background, but our views both share the emphasis on the importance of basic research – looked like this: as a PhD student of Fritz Melchers, who has been permanent member of the Basel Institute of Immunology for 15 years and is now its director, it was my task to study the heterogeneity of antibodies to an enzyme antigen. The most important results (which I collected during the last half year of my thesis time – I say this for the benefit of possible PhD students in the audience, who may be desperate about the failure of their initial experiments) were that a protein antigen may have five immunodominant regions against which antibodies can be made, that about 1,000 different antibodies are made against one of these regions, and that a mouse can make about 10 million different antibodies. These numbers had not been entirely novel at the time, but since I found them myself I now believed them. You can imagine antibodies as tiny protein rods which have two sticky ends by which they can attach to substances foreign to the body, for example bacteria or viruses or, as mentioned before, an enzyme.

During the time of my thesis 1971–1974, controversial discussions were going on in our institute whether all of these 10 million different antibodies were genetically determined and inherited, or if only a basic stock of antibody genes were inherited and the majority arise only in antibody-producing B cells by mutations (as postulated by Niels Jerne). The possibility to derive 10 or perhaps 100 different functional products from a single inherited antibody gene was so fascinating to me that I made up my mind to investigate mutations which alter the attachment sites of antibodies in Cesar Milstein's laboratory in Cambridge. This was not possible with normal antibody-producing B cells because they died away too quickly in tissue culture. However, there was a large number of B cell tumor lines, which Mike Potter in the US usually propagated by transplantation in mice, but several grew also in tissue culture. C. Milstein had a myeloma line that grew particularly well in culture and he had already found the first mutations in its antibody genes. The problem was that this line did not produce antibodies with specificity for a known antigen so that it was difficult to determine the influence of mutations on the attachment site of antibodies. I tried in C. Milstein's laboratory to make another tumor cell line, which had a known specificity, grow in tissue culture, without success. I then turned to other experiments which had to do with the fusion of two antibody-producing cell lines, hoping to learn something about the regulation of antibody synthesis.The cell hybridization experiments went unexpectedly well. As you perhaps know, it was the combination of the knowledge how to make cell hybrids with the wish to be able to grow specific antibody-producing cells in culture which led to the idea to fuse antibody-producing cells of an immunized mouse with culture-adapted tumor cells, to combine both desired properties: growth in tissue culture and production of specific antibodies.

The idea was exceedingly successful. By now dozens of companies produce monoclonal antibodies by the cell hybridization procedure. Thanks to Theo Staehelin, who moved with his research group from the institute to the company, Hoffmann-LaRoche/Basel was among the first companies which succeeded in using the technique for the characterization and purification of interferons. The global sales, which can be achieved with monoclonal antibodies, is estimated to reach 1 billion Swiss Francs in 1987.

In this way, the plan to learn something about somatic mutations of antibody genes turned into a multimillion business. It shows, like many other examples before, that basic research is the stake to win the jackpot.

The next point which I want to explain is that prize-winning work is only the tip of an iceberg of previous scientific research of many investigators. I will demonstrate this with a slide that I borrowed from Theo Staehelin. It shows schematically what is required for the production of monoclonal antibodies.

| | |
|---|---|
| 1. Immunization with antigen | – What serves as antigen?<br>– adjuvants for augmentation<br>– Sheep erythrocytes are good antigens |
| 2. Properties of immune responses | – mainly in spleen, lymph node<br>– B cells, one cell makes one antibody<br>– clonal selection (Burnett, Jerne)<br>– knowledge of antibody structure |
| 3. Existence of myeloma cell lines and experience with culture conditions | |
| 4. Knowledge about cell fusion | – Barski, Ephrussi, Harris<br>– Sendai Virus (Okada)<br>– biochemical selection of hybrid cells (Szybalski, Littlefield) |
| 5. Tests of antibody specificity | – RIA, ELISA<br>– cytotoxicity<br>– agglutination and lysis of sheep erythrocytes |

Basically the hybridoma technique is only a combination of previously known procedures. Even the goal to make monoclonal antibodies had been a dream of many immunologists for a long time. So why had nobody invented this technique before?

This is the third point which I want to elaborate on. Apparently there are problems of communication among immunologists and cell biologists. It took 10 years until (by the way, by C. Milstein and R. Cotton) cell fusion was exploited with the aim to clarify immunological questions. Unfortunately the rumor that myeloma cells are hard to fuse got confirmed in their experiments. Cell biologists had been very excited that they had learned to fuse cells of different origin to hybrid cells. They had learned early to compare the differentiation properties of the parental cells with that of the hybrid and so to learn something about regulation of differentiation. They used the properties of the hybrid cells to quite rapidly lose chromosomes for the purpose of gene mapping. For these applications it was necessary to hybridize cells of different differentiation states and of different animal species. A widespread cell fusion euphoria arose: the more distant the relationships among the parental cells, the more interesting it seemed to fuse them to hybrid cells and to find out which properties the hybrids might have. This approach was also the reason for the failure of two American investigators, Schwaber and Cohn, who about 1 year before Köhler and Milstein fused mouse myeloma cells with human antibody-producing cells with the aim to make human monoclonal antibodies. The interspecies hybrids rapidly lost human chromosomes and thus the capacity to produce antibodies: they were unstable. To obtain stable antibody-producing hybrids it was not only necessary to use tumor cells in the same state of differentiation but also of the same species as fusion partner for the antibody-producing but mortal body cell.

This type of fusion was against the common trend and had thus apparently been overlooked by the fusion experts. This is not to blame the experts. But it shows how interesting and important directions of research become fashionable and lose creativity to some extent. Nowadays the discovery and characterization of the so-called oncogenes seem similarly "fashionable" to me.

The fourth and last point I want to make is to point out how much luck is involved in appearing on the tip of an iceberg in natural sciences. The hybridoma technique is relatively simple to master, simple because two unexpected facts fell together in its lap. Antibody-producing cells of a given specificity are rare in the spleen even after immunization (about one per 1,000) and only few of these infrequent cells produce antibodies in maximal quantity. In the population of hybrids antibody-producing cells are 10- to 50-fold enriched, because recently stimulated dividing cells have an advantage in fusion. All hybridoma cells produce large amounts of antibody, a property which is forced upon them by the fusion partner. These two properties were instrumental in the rapid acceptance of the hybridoma technique worldwide.

Let me shortly summarize the points which occurred to me in connection with the hybridoma technique:

1. The hybridoma technique is a perfect example how rapidly and unexpectedly basic research can lead to profitable application. Financing organizations forget this too readily.
2. Prize-winning results are only the dot on the i of a long scientific development. This should be stressed more often also in the public press.
3. The development of new ideas gets slowed down by trends of fashion in research. Therefore it is necessary to provide young inexperienced and therefore less influenced investigators with work opportunities. The Basel Institute has established this system by the annual exchange of 10–15 of its about 50 mostly young investigators, and seems to profit from it.

In about 6 weeks I will resume my position at the Max-Planck Institute for Immunobiology in Freiburg. Therefore allow me to wish the Basel Institute of Immunology and all other research facilities in Basel all my best and a long existence, and one or the other of the researchers here the bit of luck that I had in getting this outstanding prize.

## Science forum of the Social Democratic Party of Germany, Bonn

16 October 1992

Ladies and gentlemen,
With a population density of 6 billion on earth we need science and technology to generate sufficient quantities of food, to effectively combat epidemics, to keep ourselves warm, and to protect ourselves against adverse environmental conditions. Modern science is a relatively young invention of men, perhaps 200–300 years old. With the recognition and definition of the rules and laws of nature science has made goal-directed experimentation possible. In this way, the speed by which profitable inventions are made has dramatically increased.

It is useless to ask whether by improved agriculture, by improved hygiene, and by elimination of plagues, science has contributed to today's population explosion, and has thus made itself indispensable. Except for a reduction in population density – here the churches and politics are in demand – it were the sciences that have found decent solutions for famine and plagues. It is often forgotten that the global crop production has doubled between 1950 and 1970, and that this has been possible through a combination of new plant species with improved insecticides and improved fertilizers. In "Gulliver's Travels" Jonathan Swift reports about the king of Brobdingnang: he declared, he who can make two cones of maize or two blades of grass grow on a lot of ground where hitherto only one would grow, would do more for mankind and do a greater service to his land than all politicians together.

Modern biology has with the help of gene technology, entirely new opportunities to increase agricultural plant produce. Resistances against diseases which could so far only be achieved by crossing with usually more resistant wild-type plants, an endeavor that may take decades, can now be generated in much less time by insertion of resistance genes. Examples are increased resistance in tobacco against tobacco mosaic virus or in potato against Y-virus, by introduction of the genes for the appropriate coat proteins. Other applications aim at avoiding the feeding of insect larvae by inserting appropriate insecticide genes into the plant. It is hard to rationalize why one should refrain from bringing such plants out in the field. More critical are certainly the attempts at using herbicide resistances, if the only aim of these measures is that higher concentrations of herbicide (since the recombinant plant is now resistant) can be distributed on the field. The possibilities to make agricultural plants resistant against easily biodegradable and thus ecologically tolerable substances is a legitimate goal. It is only to be taken care that the genetically altered plant does not become toxic, and maintains its taste.

There are arguments that genes can get transferred into other species or into related plants that grow naturally in the field. These arguments are not to be dismissed. I think this will indeed take place. It is not known how frequent this will be but it will happen, by the same mechanisms that operate already today. So far they have caused us no problems. Why should we then worry if this occurs with transplanted genes?

Like physics that has given us radioactive pollution, or chemistry that has caused chemical pollution of the soil, water, and air, biology will contribute to gene pollution on earth. We do not know how fast genes can get transferred within one species or how fast genes can overcome species barriers. I am therefore in favor of the registration of all cases in which genes are brought out in the field. With microbial species whose spread is hard to control, such as fungi, bacteria, and viruses, the conditions for approval must be more strict than with agricultural plants. Nevertheless, the transplanted genes cross species barriers in the same way as normal genes. Anxiety of gene liberation is inappropriate. However, I think it is reasonable to provide later generations with the opportunity to distinguish the genes that had been transplanted by the human hand from those with which evolution has operated so far. It is possible to insert short DNA sequences in each transplanted gene which facilitate the identification of a recombinant gene. One could go as far as to design the recognition sequences in the way of license plates of cars, so that a recombinant gene can be traced back to its manufacturer.

Let me introduce as another example the stunning development which the production of monoclonal antibodies, a topic with which I am very familiar, has made by the application of gene technology. The impression I want to confer to you is how lively, ingenious, and progressive gene technology can influence a certain area of research, and that basic research of biology and medicine is unthinkable today without gene technology.

For the production of monoclonal antibodies a mouse has to be immunized. A few of the spleen cells then produce the desired antibody against the immunogen. But in cell culture systems the antibody-producing lymphocytes will die off. The trick is now to keep them alive as hybrid cells by cell hybridization with a culture-adapted tumor cell. The hybrid cells grow in culture medium and produce the desired antibody. There are millions and more different antibodies and each one of them we can produce by this method, if we so wish.

When we introduced this method in the scientific journal *Nature* in 1975, the last sentence read: "Such cultures could be valuable for medical and industrial use." Cesar Milstein and I myself have considered the possibility that this sounded too boasting – today I think that without this sentence one would have doubted our ability to correctly see the implications of our discovery. Out of a project of basic research (topic: somatic mutations of antibody genes) mutimillion dollar sales are being made today with this technique. Financing institutions readily forget that support of basic research is a stake which leads to big revenues. Conversely, researchers readily forget that applied research rarely leads to scientific breakthroughs and that aiming at the question "how to make extra money with my research?" is not a suitable approach to fundamental discoveries.

For several years gene technology, which was at its very beginnings in 1975, had no influence on the method of making monoclonal antibodies. Already by the late 1980s this changed drastically. Since for the purpose of therapy antibodies of human and not of animal origin were required, and since the cell hybridization procedure did not work optimally with human cells for a variety of reasons, antibodies were "humanized" by gene technology. The short DNA segments from a rat antibody that were responsible for antigen contact were inserted into the corresponding human antibody genes. Against all expectations and to the surprise of the experts this microtransplantation worked. Today, such humanized antibodies are in use to avoid transplant rejection.

Antibodies consist of two protein chains, a longer and a shorter one. Both chains together bind the antigen. Antibody chains produced in bacteria had difficulties finding one another. The solution to this problem was to make single chain antibodies. The antigen binding protein domains of both chains were genetically linked to each other, so that they would not have to search for each other.

Another further development along these lines is the coupling of two binding domains leading to the production of "Janusins", in which one arm carries an antibody-derived binding domain and the other, for example, the docking site (CD4) for the human immunodeficiency virus (HIV) coat protein termed GP120. The idea behind this construct was that HIV-infected cells, which are recognizable by the expression of GP120 on their surface, can be bound by the CD4 arm and that by the other arm of the Janusin T cells can be accumulated on the virus-infected cell to destroy them. In the Janusins, for the first time, a recognition structure of non-antibody origin was coupled to an antibody binding site.

Today successful attempts are made to express in bacteriophage antibody-derived binding domains which are diversified by gene technology, in order to generate phage banks in which antibodies against any possible antigen are present in a prefabricated fashion. The hope is to eventually replace outdated methods such as antibody production by hybridization. This is so far not the case but I think that with such examples I could illustrate to you the enormous dynamic with which gene technology impacts on all biological research areas.

I want to explain to you in detail the methods which today are in use to generate transgenic animals. I know, you do not want to know all the details, but I want to be quite precise because here two technologies are combined which are both controversially discussed in the public: gene technology and manipulation of embryos. It seems important to me that you know how transgenic animals are made and how useful they are in research. Again I will stay in my specialty, immunology, to illustrate the importance of transgenic animals, again primarily mice.

In order to obtain transgenic mice one has to inject the DNA in question into a recently fertilized egg. In our laboratory this is mostly DNA that regulates the expression of an antibody. In order to obtain recently fertilized eggs, male and female mice are allowed to mate one day before. The fertilized female is killed and 5–10 eggs are washed out of the tubes. After injection of the DNA about 20 eggs are implanted surgically into the tubes of a pseudopregnant nurse. The nurse gives birth to 6–8 pups, one or two of them carry the implanted transgene in their genome and pass it on to their own offspring. In this way a new mouse strain is generated which is distinct from the original strain by one or several genes.

With the help of transgenic mouse strains the function of genes and their products can be studied in the complete animal. Our laboratory in Freiburg can no longer be imagined without transgenic mice. In immunological research we fight with the enormous diversity of – in our case – antibody-producing cells, making it impossible to trace the fate of a single cell. In transgenic mice which express a given antibody ideally all antibody-producing cells make the same antibody. Only now can we ask the important questions in immunobiology.

How, for example, does the body deal with cells which produce antibodies to antigens which are present in the body itself? A problem which normal animals have to solve all the time, because otherwise the immune system would destroy the body's own tissues instead of protect against pathogenic microbes. Today we know, with the help of the simplified immune systems of transgenic mice, in detail how autoaggressive immunocytes are kept in check and ultimately eliminated. This knowledge contributes to our better understanding of human autoimmune diseases and eventually perhaps to more rational therapies.

An even more important technology for the production of transgenic animals rests on the use of embryonic stem cells. These are derived of an early embryonic stage, the blastula. At this stage the embryonic stem cells which later develop into the embryo proper are surrounded by the cells which later develop into the placenta. Embryonic stem cells can grow in tissue culture, can be frozen down and thawed up again, and they can be genetically altered and still contribute to a complete embryo, if reinjected into a blastocyst. If the embryonic stem cell was derived from a mouse strain with brown coat color and if it was injected into a blastocyst of a mouse strain with white coat color, this gives rise to brown-white spotted mice which not only in their fur coat but in all other organs as well consist of an unpredictable mixture of two embryos. You may have come across such an animal in the public press, made of embryonic stem cells of a sheep and a goat, dubbed as "shoat".

Because embryonic stem cells grow in tissue culture it is relatively easy to introduce foreign DNA. You just have to bathe them in the DNA to be introduced. By an electrical impulse which transiently opens the cell membrane the foreign DNA enters the cell and inserts itself at some position into its chromosomes. In this way it is possible to trace the very rare events in which the insertion of the introduced gene – in this case a mutated form of a wild-type gene of the mouse – has taken place at the very same position – the same 0.01 mm of the about 1 m length of the string of DNA – at which the wild-type gene is found. In this case the normal gene gets replaced by the non-functional mutant gene. In animals derived from embryonic stem cells treated in this way one can study which role a particular gene has in the complicated interplay of genes and cells in the complete organism. It is often surprising that the results derived from cell cultures in petri dishes are overestimated in their importance for the whole organism. In the body the possibility exists to compensate for the loss of a gene product. In many cases, however, defects can be observed which often provide new and so far unknown conclusions about the function of the corresponding gene product. Similarities of the deficiencies in the animal to those found in heritable human diseases can help shorten the lengthy search for the defective gene in patients.

It cannot be left out to say a few words about animal protection at this point. For many scientists it was quite sobering to learn that legislature has denied them the right to judge in their own responsibility the importance of the aims that they pursue by animal experimentation. According to the animal protection law of 1986 this is now the task of a "committee" which reports to the supervising agency. One-third of the committee members are candidates nominated by the animal protection agency, two-thirds are biologists and physicians. Research institutes cannot nominate. They have to appoint an animal protection officer, scientists have to write protocols.

For a scientist, like for any other citizen in our society, it is important to live in harmony with his environment. The researchers addressed by the law keep to the regulations and secretly hope for stable conditions so as to find more time for their research which society demands of them. Far from the truth! There is no peace. Only in May a law initiative of members of the social democratic fraction has been rejected by parliament. At present the parliament discusses novel initiatives to the animal protection law by some state governments. In this context every single one is requested to take a clear position for the requirements of science. The issue is not only the money – conditions for research in Germany as a whole are at stake.

## Appendix B

# Prizes and awards to Georges Köhler

Prize for Cell Biology, Federal Ministry of Science and Education, 1978.
Albion O. Bernstein, M.D. Prize, New York Medical Society, 1980.
Gairdner Foundation Prize, Toronto, 1981.
Prize Biochemical Analytics, German Society for Clinical Chemistry, 1982.
David Pressman Memorial Award, 1982.
Landsteiner-Award, American Association of Blood Banks, 1982.
Dr. Saal van Zwangenberg-Lecture, University of Leiden, The Netherlands, 1983.
Philip Levine Prize, American Society of Clinical Pathologists, 1983.
Honorary Doctorate, Limburgs Universitair Centrum, Belgium, 1983.
Honorary Docentship, University of Basel, 1984.
Wilhelm Warner Prize, Medizinische Gesellschaft, Hamburg, 1984.
John Scott Medal, City of Philadelphia, 1984.
Albert Lasker Award for Medical Research, 1984.
Nobel Prize for Physiology or Medicine, Karolinska Institute, Stockholm, 1984.
Carus Medal, Deutsche Akademie der Naturforscher Leopoldina, 1985.
Honorary Member, International Academy of Sciences, 1985.
Honorary Member, American Association of Immunologists, 1985.
Member, European Academy of Art, Science, and Humanity, 1985.
Member, Deutsche Akademie der Naturforscher Leopoldina, 1985.
The UMKC-Visiting Professor-Award, University of Missouri, Kansas City, 1986.
Bavarian Maximilian's Order for Science and Arts, 1986.
Honorary Professorship, Beijing Medical University, 1986.
Member, Academia Europaea, Cambridge, UK. 1989.

# References and sources

## Chapter 1

Kindt, T. Capra, J.D. 1984. The antibody enigma. Plenum Press, New York, London

Von Behring, E.A., Kitasato, S. 1890. Über das Zustandekommen der Diphterieimmunität bei Tieren. Dtsch. Med. Wochenschrift. 16:1113.

Bäumler, E. 1984. Paul Ehrlich, scientist for life. Holmes and Meier Publ. New York, London.

Ehrlich, P. 1897. Zur Kenntnis der Antitoxinwirkung. Fortschr. D. Med. 15: 41.

Ehrlich, P. 1906. On immunity with special reference to cell life. Proc. R. Soc. London. (Biol.). 66:424.

Landsteiner, K. 1947. The specificity of serological reactions. Harvard University Press, Cambridge, Massachusetts.

Avery, O.T., MacLeod, C.M., McCarty, M. 1944. Studies on the chemical nature of the substance inducing transformation of pneumococcal types: Induction of transformation by desoxyribonucleic acid fraction isolated from pneumococcal Type III. J. Clin. Invest. 179: 137.

Watson, J. 1966. Molecular biology of thje gene. W.A. Benjamin Inc. Publ. New York.

Alberts, B., Bray, D., Lewis, J., Raff, M., Roberts, K., Watson, J.D. 1983. Molecular biology of the cell. Garland Publ. New York, London.

Watson, J.D., Crick, F.H. 2003 (reprinted). A structure for deoxyribose nucleic acid. 1953. Nature 421: 397.

Crick, F.H.C. 1966. The genetic code. Sci. Am. 215:55.

Edelman, G.M., Gally, J.A. 1964. A model for the 7S antibody molecule. Proc. Nat. Acad. Sci. USA. 51: 846.

Pauling, L. 1940. A theory of the structure and process of the formation of antibody. J. Am. Chem. Soc. 62:2643.

Breinl, F., Haurowitz, F. 1930. Chemische Untersuchung das Präzipitates aus Hämoglobin und Anti-hämoglobin-serum und Bemerkungen über die Natur der Antikörper. Z. Phys. Chem. 192:45.

Jerne, N.K. 1955. The natural selection theory of antibody formation. Proc. Nat. Acad. Sci. USA. 41:849.

Burnet, F.M. 1957. A modification of Jerne's theory of antibody production using the concept of clonal selection. Aust. J. Sci. 20:67.

Lederberg, J. 1959. Genes and antibodies. Do antigens bear instructions for antibody specificity or do they select cell lines that arise by mutation? Science 152:1513.

Kunkel, H.G. 1965. Myeloma proteins and antibodies. Harvey Lect. 59:219.

Kunkel, H.G. 1970. Individual antigenic specificity, cross specificity, and diversity of human antibodies. Fed. Proc. 29:55.

Natvig, J.G., Kunkel, H.G. 1973. Human immunoglobulins: Classes, subclasses, genetic variants, and idiotypes. Adv. Immunol. 16:1

Hilschmann, N., Craig, L.C. 1965. Amino acid sequence studies with Bence-Jones proteins. Proc. Nat. Acad, Sci. USA. 53:1403.

Dreyer, W.N., Bennett, J.C. 1965. The molecular basis of antibody formation: A paradox. Proc. Nat. Acad. Sci. USA. 54:409.

Hood, L., Talmadge, D.W. 1970. Mechanism of antibody diversity: Germline basis for variability. Science 168:325.

Cohn, M. 1970. Selection under a somatic model. Cell. Immunol. 1: 461.

Doolittle, W.F. 1978. Genes in pieces: Where they ever together? Nature 272: 581.

Ohno, S. 1970. Evolution by gene duplication. Springer Verlag, New York.

Gilbert. W. 1979. Introns and exons: Playgrounds of evolution. In: Eucaryotic gene regulation, Eds: Axel, R., Maniatis, T., Fox, C.F. Academic Press.

Darnell, J.E. 1978. Implications of RNA – RNA splicing in evolution of eukaryotic cells. Science 202:1257.

Hozumi, N., Tonegawa, S. 1976. Evidence for somatic rearrangement of immunoglobulin genes coding for variable and constant regions. Proc. Nat. Acad. Sci. USA. 73:3632.

Tonegawa, S., Maxam, A.M., Tizard, R., Bernard, O., Gilbert, W. 1978. Sequence of a mouse germ-line gene for a variable region of an immunoglobulin light chain. Proc. Nat. Acad. Sci. USA. 75: 1485.

Gerhart, P. 1882. Generation of immunoglobulin variable gene diversity. Immunol. Today 3: 107.

## Chapter 2

Söderquist, T. 2003. Science as autobiography, the troubled life of Niels Jerne. Yale University Press, New Haven, London.

Jerne, N. 1951. A study of avidity based on rabbit skin responses to diphteria toxin-antitoxin mixtures. Munksgaard, Copenhagen

Jerne, N. 1956. The presence in normal serum of specific antibody against bacteriophage T4 and its increase during the earliest stages of immunization. J. Immunol. 76: 209.

Jerne, N.K. 1955. The natural selection theory of antibody formation. Proc. Nat. Acad. Sci. USA. 41:849.

Burnet, F.M. 1957. A modification of Jerne's theory of antibody production using the concept of clonal selection. Aust. J. Sci. 20:67.

Burnet, F. M. 1959. The clonal selection theory of acquired immunity. Cambridge University Press, Cambridge.

Pauling, L. 1940. A theory of the structure and process of the formation of antibody. J. Am. Chem. Soc. 62:2643.

Breinl, F., Haurowitz, F. 1930. Chemische Untersuchung das Präzipitates aus Hämoglobin und Anti-hämoglobin-serum und Bemerkungen über die Natur der Antikörper. Z. Phys. Chem. 192:45.

Lederberg, J. 1959. Genes and antibodies. Do antigens bear instructions for antibody specificity or do they select cell lines that arise by mutation? Science 152:1513.

Ehrlich, P. 1906. On immunity with special reference to cell life. Proc. R. Soc. London. (Biol.). 66:424.

Nossal, G.J.V. 1958. Antibody production by single cells. Brit. J. Exp. Physiol. 39:544.
Jerne, H.K., Nordin, A.A. 1963. Plaque formation in agar by single antibody-producing cells. Science 140: 405.
Jerne, N. 1971. The somatic generation of immune recognition. Eur. J. Immunol. 1:1
Simonsen, M. 1967. The clonal selection hypothesis evaluated by grafted cells reacting against their host. Cold Spring Harbor Symp. Quant. Biol. 32:517.
Jerne, N. 1974. Towards a network theory of the immune system. Annales d'Immunologie 125C:373.
Oudin, J. Michel, M. 1963. Une nouvelle forme d'allotypie des globulines-γ du serum de lapin, apparemment liee a la fonction et a la specificite anticorps. Comptes Rendu des Seances de l'academies des Sciences, Ser. 3257:805.
Wigzell, H. 1985. The Nobel Prize for physiology or medicine. In: Les Prix Nobel 1984: Nobel Prizes, Presentations, Biographies and Lectures. Almquist and Wiksell International, Stockholm.
Milstein, C. 1984. From the structure of antibodies to the diversification of the immune response. In: Les Prix Nobel 1984: Nobel Prizes, Presentations, Biographies and Lectures. Almquist and Wiksell International, Stockholm.
The Official Web Site of The Nobel Foundation: http://www.nobel.se/medicine/laureates/1984/milstein
Davies, D.R., Padlan, E.H., Segal, D.M. 1975. Three dimensional structure of immunoglobulins. Annu. Rev. Biochem. 44: 639.
Hilschmann, N., Craig, L.C. 1965. Amino acid sequence studies with Bence-Jones proteins. Proc. Nat. Acad, Sci. USA. 53:1403.
Dreyer, W.N., Bennett, J.C. 1965. The molecular basis of antibody formation: A paradox. Proc. Nat. Acad. Sci. USA. 54:409.
Smithies, O. 1967, Antibody variability. Somatic recombination between the elements of "antibody gene pairs" may explain antibody variability. Science 157:276.
Milstein, C. 1966. Variations in amino-acid sequence near the disulphide bridges of Bence-Jones proteins. Nature 209:370
Brenner, S., Milstein, C. 1966. Origin of antibody variation. Nature 211:242.
Milstein, C. 1967. Linked groups of residues in immunoglobulin κ-chains. Nature 216:330.
Milstein, C., Milstein, C.P., Feinstein, A. 1969. Non-allelic nature of the basic sequences of normal immunoglobulin κ-chains. Nature 221: 151.
Milstein, C., Brownlee, G.G., Harrison, T.M., Mathews, M.B. 1972. A possible precursor of immunoglobulin light chains. Nature New Biol. 239:117.
Potter, M. 1972. Immunoglobulin-producing tumors and myeloma proteins in mice. Physiol. Rev. 52: 631.
Horibata, K., Harris, A.W. 1970. Mouse myelomas and lymphomas in culture. Exp. Cell Res. 60: 61.
Milstein, C., Adetogbo, K., Cowans, N.J., Secher, D.S. 1974. Clonal variants of myeloma cells. In: Progress in Immunology. Eds: L.Brent, J. Holborrow, 1:157. North Holland, Amsterdam.
Secher, D.S. Milstein, C. Adetugbo, K. 1977. Somatic mutants and antibody diversity. Immunol. Rev. 36: 51
Rudikoff, S., Giusti, A.M., Cook, W., Scharff, M.D. 1982. Single amino acid substitution altering antigen-binding specificity. Proc. Nat. Acad. Sci. USA. 79:1979.
Rajewsy, K., Schirrmacher, V., Nase, S., Jerne, N.K. 1969. The requirement of more than one antigenic determinant for immunogenicity. J. Exp. Med. 129:1131.

Rajewsky, K. 1971. The carrier effect and cellular cooperation in the induction of antibodies. Proc. Roy. Soc. Lond. Biol Sci. 176:385.

## Chapter 3

MPG-Archiv, II.Abt.Rep. 1A PA Köhler.
Wade, N. 1982. Hybridomas: The making of a revolution. Science 215:1073.
Köhler, G., Hanke, W. 1973. Einsatz des Computerunterstützten Unterrichts (CUU) in Tutorien und Übungen am Beispiel der Lehrveranstaltung "Einführung in die Genetik". In: Unterrichtswissenschaft, p53.
Melchers, F. 1971. Biosynthesis, transport, and secretion of immunoglobulin in plasma cells. Histochem. J. 3:389.
Rogers, J., Early, P., Carter, C., Calame, K., Bond, M., Hood, L., Wall, R. 1980. Two mRNAs with different 3'ends encode membrane-bound and secreted forms of immunoglobulin chains. Cell 20:303.
Bonner, W.A., Hulett, H.R., Sweet, R.G., Herzenberg, L.A. 1972. Fluorescence activated cell sorting. Rev. Sci. Instrum. 43: 404.
Melchers, F., Messer, W. 1973. The mechanism of activation of mutant-β-galactosidase by specific antibodies. Eur. J. Biochem. 35: 380.
Fazekas de St Groth, S., Webster, R.G. 1966. Disquisitions of original antigenic sin. I. Evidence in man. J. Exp. Med. 124: 331.
Köhler, G., Melchers, F. 1972. Isoelectric focussing of antibodies which activate mutant β-galactosidases. Eur. J. Immunol. 2: 543.
Köhler, G., 1976. Frequency of precursor cells against the enzyme β-galactosidase. An estimate of the Balb/c strain antibody repertoire. Eur. J. Immunol. 6:340.
Luria, S. E., Delbrück, M. 1943. Mutations of bacteria from virus sensitivity to virus resistance. Genestics 28:491.

## Chapter 4

Kindt, T. Capra, J.D. 1984. The antibody enigma. Plenum Press, New York, London
Edelman, G.M. 1991 (reprinted). Lecture for the Nobel Prize for physiology or medicine 1972: Antibody structure and molecular immunology. Scand. J. Immunol. 34:2.
Porter, R.R. 1991 (reprinted). Lecture for the Nobel Prize for physiology or medicine 1972: Structural studies of immunoglobulins. Scand. J. Immunol. 34: 381.
Edelman, G.M., Gally, J.A. 1964. A model for the 7S antibody molecule. Proc. Nat. Acad. Sci. USA. 51: 846.
Kunkel, H.G. 1965. Myeloma proteins and antibodies. Harvey Lect. 59:219.
Kunkel, H.G. 1970. Individual antigenic specificity, cross specificity, and diversity of human antibodies. Fed. Proc. 29:55.
Natvig, J.G., Kunkel, H.G. 1973. Human immunoglobulins: Classes, subclasses, genetic variants, and idiotypes. Adv. Immunol. 16:1
Edelmann, G.M. 1971. Antibody structure and molecular immunology. Ann. N.Y. Acad. Sci. 190:5.
Hilschmann, N., Craig, L.C. 1965. Amino acid sequence studies with Bence-Jones proteins. Proc. Nat. Acad, Sci. USA. 53:1403.
Kratzin H, Yang CY, Gotz H, Pauly E, Kolbel S, Egert G, Thinnes FP, Wernet P, Altevogt P, Hilschmann N. 1981. [Primary structure of class II human histocompatibility antigens. 1st communication. Amino acid sequence of the N-terminal

198 residues of the beta chain of a HLA-Dw2,2;DR2,2-alloantigen (author's transl)]. Hoppe Seyler's Z. Physiol. Chem. 362:1665. German.
Metzger, H. 1969. Myeloma proteins and antibodies. Am. J. Med. 47:837.
Davies, D.R., Padlan, E.H., Segal, D.M. 1975. Three dimensional structure of immunoglobulins. Annu. Rev. Biochem. 44: 639.
Potter, M. 1972. Immunoglobulin-producing tumors and myeloma proteins in mice. Physiol. Rev. 52: 631.
Krause, R.M. 1970. The search for antibodies with molecular uniformity. Adv. Immunol. 12:1.
Givas, S.J., Centeneo, E.R., Manning, M., Sehon, A.H. 1968. Isolation of antibodies to the C-terminal heptapeptide of myoglobin with a synthetic peptide. Immunochemistry. 5:314.
Haber, E. 1970. Antibodies of restricted heterogeneity for structural study. Fed. Proc. 29:66.
Nisonoff, A., MacDonald, A.B., Hopper, J.E., Daugherty, H. 1970. Quantitative studies of idiotypic antibodies. Fed. Proc. 29:72.
Roholt, O.A., Seon, B.K., Pressmann, D. 1970. Antibodies of limited heterogeneity: L chains of a single mobility. Immunochemistry. 7:329.
Brenneman, L., Singer, S.J. 1968.The generation of antihapten antibodies with electrophoretically homogeneous L chains. Proc. Natl. Acad. Sci. USA 60:258.
Eichmann, K. 1975. Genetic control of antibody specificity in the mouse. Immunogenetics. 2:491.
Wu, T.T., Kabat. E.A. 1970. An analysis of sequences of the variable regions of Bence-Jones proteins and myeloma light chains and their implications for antibody complementarity. J. Exp. Med. 132:211.
Capra, J.D., Edmundson, A.B. 1977. The antibody combining site. Sci. Am. 236:50.
Eichmann, K., Berek. C. 1973. Mendelian segregation of a mouse antibody idiotype. Eur. J. Immunol. 3:599.
Eichmann, K., Tung, A.S., Nisonoff A. 1974. Linkage and rearrangement of genes encoding mouse immunoglobulin heacy chains. Nature 250:509.
Jerne, N.K. 1980, Forword. In: Idiotypes, what they said at the time. Ed. Schnurr, I. The Basel Institute of Immunology, Publ.
Rabbits, T., Milstein, C. 1975. Mouse immunoglobulin genes. Studies on the reiteration frequency of light chain genes by hybridization procedure. Eur. J. Biochem. 52:125.
Tonegawa, S., Bernardini, A., Weimann, B.J., Steinberg, C. 1974. Reiteration frequency of antibody genes. FEBS Lett. 40:92.
Leder, P. 1982. The genetics of antibody diversity. Sci. Am. 246: 108.

## Chapter 5

Barski, G., Sorieul, S., Cornefert, M. 1960. Production dans les cultures in vitro de deux souches cellulaire en association, de cellules de caractere "hybride". Compt. Rend. Acad, Sci. Paris. 251:1825.
Sorieul, S., Ephrussi, B. 1961, Karyological demonstration of hybridization of mammalian cells in vitro. Nature 190: 653.
Ephrussi, B., Sorieul, S. 1962. Mating of somatic cells in vitro. In: Approaches to the genetic analysis of mammalian cells. Eds: Merchant, D.J. Neel. J.V. Univ. of Mitchgan Press, p 81.

Gershon, D., Sachs, L. 1963. Properties of a somatic hybrid between mouse cells of different genotype. Nature 198:912.
Okada, Y. 1962. Analysis of giant polynuclear cell formation caused by HVJ virus from Ehrlich's ascites tumor cells. I. Microscopic observation of giant polynuclear cell formation. Exp. Cell. Res. 26:98.
Roizman, B. 1962. Polykaryocytosis induced by viruses. Proc. Natl. Acad. Sci. USA. 48:228.
Harris, H., Watkins, J.F. 1965. Hybrid cells derived from mouse and man: Artificial heterokaryons of mammalian cells from different species. Nature 206: 640.
Harris, H. 1970. The use of cell fusion in the analysis of gene action. Proc. R. Soc. Lond. B Biol. Sci. 176:315.
Littlefield, J.W. 1964. Selection of hybrids from matings of fibroblasts in vitro and their presumed recombinants. Science 145:709.
Szybalski, W. 1958. Resistanve to 8-azaguanine, a selective genetic marker for a human cell line. Microbiol. Genetics Bull. 16:30.
Szybalski, W., Szybalska, E.H., Ragni, G. 1962. Genetic studies with human cell lines. National Cancer Inst. Monograph 7: 75.
Kit, S., Dubbs, D.R., Piekarski, L.J., Hsu, T.C. 1963. Deletion of thymidine kinase activity from L cells resistant to bromodeoxyuridine. Exp. Cell Res. 31:297.
MPG-Archiv, ZA 143 Köhler, AO69.
Periman, P. 1970. IgG synthesis in hybrid cells from an antibody-producing mouse myeloma and an L cell substrain. Nature 228:1086.
Coffino, P., Knowles, B., Nathensen, S.G., Scharff, M.D. 1971. Suppression of immunoglobulin synthesis by cellular hybridization. Nat. New. Biol. 231:87.
Bevan, M.J., Parkhouse, R.M., Williamson, A.R., Askonas, B.A. 1972. Biosynthesis of immunoglobulins. Prog. Biophys. Mol. Biol. 25:133.
Mohit, B. Fan, K. 1971. Hybrid cell line from a cloned immunoglobulin-producing mouse myeloma and a nonproducing mouse lymphoma. Science 171: 75.
Pernis, B., Chiappino, G., Kelus, A., Gell, P.G.H. 1965. Cellular localization of immunoglobulins with different allotypic specificities in rabbit lymphoid tissue. J. Exp. Med. 122: 853.
Weiler, E. 1965. Differential activity of allelic γ-globulin genes in antibody-producing cells. Proc. Nat. Acad. Sci. USA. 54:1765.
Bazin, H., Deckers, C., Beckers, A., Heremanns, J.F. 1972. Transplantable immunoglobulin-secreting tumours in rats. I. General features of LOU-Ws1 strain rat immunocytomas and their monoclonal proteins. Int. J. Cancer. 10:568.
Cotton, R.G.H., Milstein. C. 1973. Fusion of two immunoglobulin-producing myeloma cells. Nature 244: 42.
Schwaber, J., Cohen, E.P. 1974. Pattern of immunoglobulin synthesis and assembly in a human-mouse somatic cell hybrid clone. Proc. Nat. Acad. Sci. USA. 71:2203.

## Chapter 6

Milstein, C. 1984. From the structure of antibodies to the diversification of the immune response. In: Les Prix Nobel 1984: Nobel Prizes, Presentations, Biographies and Lectures. Almquist and Wiksell International, Stockholm.
Potter, M. 1972. Immunoglobulin-producing tumors and myeloma proteins in mice. Physiol. Rev. 52: 631.
Cohen, S., Milstein, C. 1967. Structure and biological properties of immunoglobulins. Adv. Immunol. 7:1.

Brownlee, G.G., Cartwright, E.M., Cowan, N.J., Jarvis, J.M., Milstein, C. 1973. Purification and sequence of messenger RNA for immunoglobulin light chains. Nat. New Biol. 244:236.
Secher, D.S., Cotton, R.G., Milstein, C. 1973. Spontaneous mutation in tissue culture-chemical nature of variant immunoglobulin from mutant clones of MOPC 21. FEBS Lett. 37:311.
Milstein, C., Adetogbo, K., Cowans, N.J., Secher, D. S. 1974. Clonal variants of myeloma cells. In: Progress in Immunology. Eds: L.Brent, J. Holborrow, 1:157. North Holland, Amsterdam.
Secher, D.S. Milstein, C. Adetugbo, K. 1977. Somatic mutants and antibody diversity. Immunol. Rev. 36: 51.
Wade, N. 1982. Hybridomas: The making of a revolution. Science 215:1073.
Metzger, H. 1969. Myeloma proteins and antibodies. Am. J. Med. 47:837.
Eisen, H.N., Michaelides, M.C., Underdown, B.J., Schulenburg, E.P., Simms, E.S. Experimental approaches to homogenous antibody populations. Myeloma proteins with antihapten antibody activity. Fed. Proc. 29:78.
MPG-Archiv, ZA 143 Köhler, AO69.
Deutsches Museum Bonn, Köhler's Laborbuch.
Askonas, I., Bard, B., Booth, C., Bud, R., Gallowy, J., Gowans, J., Gray, J., Köhler, G., Milstein, C., Newell, J., Owen, D., Secher, D., Tyrell, D., Winchester, G., Weatherall, M., Williams, P. 1997. Wellcome Witnesses to Twentieth Century Medicine. Technology transfer in Britain: The case of monoclonal antibodies. 1993. Eds: Tansey, E.M., Catteral, P.P. Wellcome Institute for the History of Medicine. 1:1.
Rajewsky, K. 2002. Obituary: Cesar Milstein (1927–2002). 416:806.
Alkan, S.S. 2004. Monoclonal antibodies: the story of a discovery that revolutionized science and medicine. Nature Rev. Immunol. 4:153.
Karpas, A. 2004. Cesar Milstein and the discovery of monoclonal antibodies. Nature Rev. Immunol. Publ. Online Nov. 1. 2004.
Cunningham, A.J., Pilarski, L.M. 1974. The generation of antibody diversity. II. Plaque morphology as a simple marker for antibody specificity at the single-cell level. Eur. J. Immunol. 4:757.
Arlett, C.F., Turnball, D., Harcourt, S.A., Lehmann, A.R., Colella, C.M. 1975. A comparison of the 8-azaguanine and ouabain-resistance systems for the selection of induced mutant Chinese hamster cells. Mutat. Res. 33:261.
Köhler, G., Milstein, C. 1975. Continuous cultures of fused cells secreting antibody of predefined specificity. Nature 256:495.
Köhler, G., Howe, S.C., Milstein, C. 1976. Fusion between immunoglobulin-secreting and nonsecreting myeloma cell lines. Eur. J. Immunol. 6:292.
Köhler, G., Milstein, C. 1976. Derivation of specific antibody-producing tissue culture and tumor lines by cell fusion. Eur. J. Immunol. 6:511.
Köhler, G., Shulman, M.J. Cellular and molecular restrictions of the lymphocyte fusion. Curr. Top. Microbiol. Immunol. 81:143.
Lemke, H., Hammerling, G.J., Hohmann, C., Rajewsky, K. 1978. Hybrid cell lines secreting monoclonal antibody specific for major histocompatibility antigens of the mouse. Nature 271:249.
Köhler, G., Pearson, C., Milstein, C. 1977. Fusion of T and B cells. Som. Cell Genetics 3:303.
Milstein, C., Adetugbo, K., Cowan, N.J. Köhler, K., Secher, D,S., Wilde, C.D. 1977.

Somatic cell genetics of antibody-secreting cells: studies of clonal diversification and analysis by cell fusion. Cold Spring Harbor Symp. Quant. Biol. 41:793.
Shulman, M.J. Wilde, D., Köhler, G. 1978. A better cell line for making hybridomas secreting specific antibodies. Nature 276:269.
Kearny, J.F., Radbruch, A., Liesegang, B., Rajesky, K. 1979. A new mouse myeloma cell line that has lost immunoglobulin expression but permits the construction of antibody-secreting hybrid cell lines. J. Immunol. 123:1548.
Köhler, G., Hengartner, H. Shulman, M.J. 1978. Immunoglobulin production by lymphocyte hybridomas. Eur. J. Immunol. 8:82.
Köhler, G. 1980. Immunoglobulin chain loss in hybridoma lines. Proc. Nat. Acad. Sci. USA. 77:2197.

## Chapter 7

Köhler, G. 1984. Derivation and diversification of monoclonal antibodies. In: Les Prix Nobel 1984: Nobel Prizes, Presentations, Biographies and Lectures. Almquist and Wiksell International, Stockholm.
Köhler, G., Shulman, M.J. 1980. Immunoglobulin M mutants. Eur. J. Immunol. 10:467.
Köhler, G. 1980. Immunoglobulin chain loss in hybridoma lines. Proc. Nat. Acad. Sci. USA 77:2197.
Melchers, F., Rolink, T., Grawunder, U., Winkler, T.H., Karasuyama, H., Ghia, P., Anderson, J. 1995. Positive and negative selection events during B lymphopoiesis. Curr. Opin. Immunol. 7:214.
MPG-Archiv, ZA 143 Köhler, LB71.
Hozumi, N., Tonegawa, S. 1976. Evidence for somatic rearrangement of immunoglobulin genes coding for variable and constant regions. Proc. Nat. Acad. Sci. USA. 73:3628.
Tonegawa ,S,. Maxam, A.M., Tizard, R., Bernard, O., Gilbert, W. 1978. Sequence of a mouse germ-line gene for a variable region of an immunoglobulin light chain. Proc. Nat. Acad. Sci. USA 75:1485.
Weigert, M.G., Cesari, I.M,. Yonkovich, S.J., Cohn, M. 1970. Variability in the lambda light chain sequences of mouse antibody. Nature 228:1045.
Köhler, G., Potash, M.J., Lehrach, H., Shulman, M.J. 1982. Deletions in immunoglobulin mu chains. EMBO J. 1:555.
Shulman, M.J., Heusser, C., Filkin, C., Köhler, G. 1982. Mutations affecting the structure and function of immunoglobulin M. Mol. Cell. Biol. 2:1033.
Köhler, G., Baumann, B., Iglesias, A., McCubrey, J., Potash, M.J., Traunecker, A., Zhu, D. 1984. Different ways to modify monoclonal antibodies. Med. Oncol. Tumor Pharmacother. 1:227.
MPG-Archiv, ZA 143 Köhler, LB106.
Sidman, C., Potash, M.J., Köhler, G. 1981. Roles of protein and carbohydrate in glycoprotein processing and secretion. Studies using mutants expressing altered IgM mu chains. J. Biol. Chem. 256:13180.
Leptin, M., Potash, M.J., Grutzmann, R., Heusser, C., Shulman. M., Köhler. G., Melchers, F. 1984. Monoclonal antibodies specific for murine IgM I. Characterization of antigenic determinants on the four constant domains of the mu heavy chain. Eur. J. Immunol. 14:534.
Mathur, A., Lynch, R.G., Köhler, G. 1988. Expression, distribution and specificity of Fc receptors for IgM on murine B cells. J. Immunol. 141:1855.

Mathur, A., Lynch, R.G., Köhler, G. 1988. The contribution of constant region domains to the binding of murine IgM to Fc mu receptors on T cells. J. Immunol. 140:143.

Iglesias, A,, Kopf, M., Williams, G.S., Buhler, B., Köhler, G. 1991. Molecular requirements for the mu-induced light chain gene rearrangement in pre-B cells. EMBO J. 10:2147.

Hamlyn, P.H., Brownlee, G.G., Cheng, C.C., Gait, M.J., Milstein, C. 1978. Complete sequence of constant and 3' noncoding regions of an immunoglobulin mRNA using the dideoxynucleotide method of RNA sequencing. Cell. 15:1067.

Griffiths, G.M., Berek, C., Kaartinen, M., Milstein, C. 1984. Somatic mutation and the maturation of immune response to 2-phenyl oxazolone. Nature 312:271.

Sablitzky, F., Wildner, G., Rajewsky, K. 1985. Somatic mutation and clonal expansion of B cells in an antigen-driven immune response. EMBO J. 4:345.

Rajewsky, K. 1989. Somatic antibody mutants. Immunol. Today 10:10.

Jacob, J., Kelsoe, G., Rajewsky, K., Weiss, U. 1991. Intraclonal generation of antibody mutants in germinal centres. Nature 354:389.

Williams, A.F., Galfre, G., Milstein, C. 1977. Analysis of cell surfaces by xenogeneic myeloma-hybrid antibodies: differentiation antigens of rat lymphocytes. Cell. 12:663.

Milstein, C., Galfre, G., Secher, D.S., Springer, T. 1979. Monoclonal antibodies and cell surface antigens. Cell. Biol. Int. Rep. 3:1.

Howard, J.C., Butcher, G.W., Galfre, G., Milstein, C., Milstein, CP. 1979. Monoclonal antibodies as tools to analyze the serological and genetic complexities of major transplantation antigens. Immunol. Rev. 47:139.

Ledbetter JA, Herzenberg LA. 1979. Xenogeneic monoclonal antibodies to mouse lymphoid differentiation antigens. Immunol. Rev. 47:63.

Herlyn, M., Steplewski, Z., Herlyn, D., Koprowski, H. 1979. Colorectal carcinoma-specific antigen: detection by means of monoclonal antibodies. Proc. Nat. Acad. Sci. USA. 76:1438.

Steplewski, Z., Herlyn, M., Herlyn, D., Clark, W.H., Koprowski H. 1979. Reactivity of monoclonal anti-melanoma antibodies with melanoma cells freshly isolated from primary and metastatic melanoma. Eur. J. Immunol. 9:94.

Wade, N. 1982. Hybridomas: The making of a revolution. Science 215:1073.

Wiktor TJ, Koprowski H. 1980. Antigenic variants of rabies virus. J. Exp. Med. 152:99.

Milstein, C., Lennox, E. 1980. The use of monoclonal antibody techniques in the study of development cell surfaces. Curr. Top. Dev. Biol. 14:1.

Peters, T. 1993. Interferon and its first clinical trial: Looking behind the scenes. Medical History 37: 270.

Secher, D.S., Burke, D.C. 1980. A monoclonal antibody for large-scale purification of human leukocyte interferon. Nature 285:446.

McMaster, W.R., Williams, A.F. 1979. Identification of Ia glycoproteins in rat thymus and purification from rat spleen. Eur. J. Immunol. 9:426.

Djerassi, C. 1989. Cantor's Dilemma. Doubleday, New York; MacDonald, London

## Chapter 8

Dosi, G. 1984. Technical change and industrial transformation. St Martins, New York.

Markle, G.E., Robin, S.S. 1985. Biotechnology and the social reconstruction of molecular biology. Science, Technology and Human Values 10: 70.
McLaren, A., Warnock, M. 1992. Our genetic future: The science and ethics of genetic technology. Oxford University Press, Oxford, UK.
Bud, R. 1993. The uses of life. A history of biotechnology. Cambridge University Press, Cambridge, UK.
Weiner, C. 1986. Universities, professors, and patents: A continuing controversy. Technology Review 89: 32.
Fjermedal. 1984. Magic bullets. Macmillan, New York.
Wade, N. 1980. Inventor of hybridoma technology failed to file for patent. Science 208: 693.
Schwartz, H. 1993. Talking points: American protectionism – the Schwartz view. Scrip Magazine June: 6.
Köhler, G., Milstein, C. 1975. Continuous cultures of fused cells secreting antibody of predefined specificity. Nature 256:495.
Koprowsky, H., Croce, C. 1980. Hybridomas revisited. Science 210: 248.
Dickson, D. 1983. Wistar denied monoclonal antibody patent in UK. Science 222: 1309.
MacKenzie, M., Cambrosio, A., Keating, 1988. The commercial application of a scientific discovery. Research Policy, 6: 36.
Askonas, I., Bard, B., Booth, C., Bud, R., Gallowy, J., Gowans, J., Gray, J., Köhler, G., Milstein, C., Newell, J., Owen, D., Secher, D., Tyrell, D., Winchester, G., Weatherall, M., Williams, P. 1997. Wellcome Witnesses to Twentieth Century Medicine. Technology transfer in Britain: The case of monoclonal antibodies. 1993. Eds: Tansey, E.M., Catteral, P.P. Wellcome Institute for the History of Medicine. 1:1.
Spinks, A. 1980. Advisory Council for Applied Research and Development/ Advisory Board of the Research Councils/Royal Society. Report of the Working Party. London HMSO.
Cambrosio, A., Keating, P. 1988. Going monoclonal: Art, science, and magic in the use of hybridoma technology. Social Problems 35: 244.
MacKenzie, M., Keating, P., Cambrosio, A. 1990. Patents and free information in biotechnology: Making monoclonal antibodies proprietary. Science, Technology, and Human Values, 15: 65.
Reinherz, E.L., Schlossman, S.F. 1980. The differentiation and function of human T lymphocytes. Cell 19:821.
Scott, M.G. 1985. Monoclonal antobodies – approaching adolescence in diagnostic immunoassays. Trends in Biotechnology 3 :170.
Klausner, A. 1985. Ortho awaits nod on therapeutic monoclonal. Bio/Technology 3: 961.
Fox, J.L. 1984. Encroaching on research freedom. Science 224: 585.
Bonner, W.A., Hulett, H.R., Sweet, R.G., Herzenberg, L.A. 1972. Fluorescence activated cell sorting. Rev. Sci. Instrum. 43: 404.

## Chapter 9

MPG-Archiv, II. Abt., Rep. 1A PA Köhler.
Henning, E., Kazemi, M. 1988. Chronik der Kaiser Wilhelm Gesellschaft zur Förderung der Wissenschaften. Veröffentlichungen aus dem Archiv der Max-Planck-Gesellschaft, Berlin.

Henning, E., Kazemi, M. 1992. Chronik der Max-Planck-Gesellschaft zur Förderung der Wissenschaften unter der Präsidentschaft Otto Hahns. Veröffentlichungen aus dem Archiv der Max-Planck-Gesellschaft, Berlin

Henning, E. 2000. Beiträge zur Wissenschaftsgeschichte Dahlems. Veröffentlichungen aus dem Archiv der Max-Planck-Gesellschaft. Berlin.

Henning, E., Kazemi, M. 2002. Dahlem – Domäne der Wissenschaft/Dahlem – Domain of Science. Veröffentlichungen aus dem Archiv der Max-Planck-Gesellschaft. Berlin.

Henning, E. 1997. Max Planck (1858–1947). Berichte und Mitteilungen der Max-Planck-Gesellschaft, München.

Heilborn, J.L. 2000. The dilemmas of an upright man. Max Planck and the fortunes of German science. Harvard University Press, Cambridge, Mass. USA.

Westphal, O. 2003. About the history of the MPI for Immunobiology. In: The Biology of Complex Organisms: Creation and Protection of Integrity. Ed: Eichmann, K. Birkhäuser Verlag, Basel.

Eichmann, K. 2004. Obituary O. Westphal. Eur. J. Immunol. 34: 3299.

Westphal, O. 1989. Immunology and the "Belle Epoche" in Berlin, adding some personal memories. In. Progr. Immunol. VII. Ed: Melchers, F. et al. p. XXXV. Springer Verlag, Berlin.

Lüderitz, O., Tanamoto, K., Galanos, C., McKenzie, G.R., Brade, H., Zahringer, U., Rietschel, E.T., Kusumoto, S., Shiba, T. 1984. Lipopolysaccharides: structural principles and biologic activities. Rev. Infect. Dis. 6:428.

Deutsche Gesellschaft für Immunologie. 1997. Festschrift 30 Jahre 1967 – 1997. Ed: Deutsche Gesellschaft für Immunologie EV. dgfi@sersys.de

Janeway, C.A.Jr. 2003. How the immune system protects the body from infection. In: The Biology of Complex Organisms: Creation and Protection of Integrity. Ed: Eichmann, K. Birkhäuser Verlag, Basel.

Eichmann, K. 2003. The second twenty years of the MPIIB. In: The Biology of Complex Organisms: Creation and Protection of Integrity. Ed: Eichmann, K. Birkhäuser Verlag, Basel.

MPG Archiv, II. Abt., 1A Biologisch-Medizinische Sektion, MPI für Immunbiologie.

## Chapter 10

MPG Archiv, ZA 159 Eichmann, I,4; III, 15-16

MPG-Archiv, II Abt., Rep. 1A PA Köhler.

MPG-Archiv, II. Abt., Rep. 1A Biologisch-Medizinische Sektion, MPI für Immunbiologie.

Köhler, G., Potash, M.J., Lehrach, H., Shulman, M.J. 1982. Deletions in immunoglobulin mu chains. EMBO J. 1:555.

Shulman, M.J., Heusser, C., Filkin, C., Köhler, G. 1982. Mutations affecting the structure and function of immunoglobulin M. Mol. Cell. Biol. 2:1033.

Secher, D.S. Milstein, C. Adetugbo, K. 1977. Somatic mutants and antibody diversity. Immunol. Rev. 36: 51.

Wade, N. 1982. Hybridomas: The making of a revolution. Science 215:1073.

Nirenberg, M.W. 1972. The genetic code. In: Nobel Lectures, Physiology or Medicine 1963–1970. Elsevier Publishing Company, Amsterdam.

Matthaei, J.H., Jones, O.W., Martin, R,G., Nirenberg, M.W. 1962. Characteristics and composition of RNA coding units. Proc. Nat. Acad. Sci. USA 48:666.

## Chapter 11

MPG Archiv, ZA 159 Eichmann, III, 15-16, 18.
MPG-Archiv, ZA 143 Köhler, AO 36-53.
MPG-Archiv, II Abt., Rep. 1A PA Köhler.
MPG Archiv, II. Abt., Rep. 1A Biologisch-Medizinische Sektion, MPI für Immunbiologie.
Riesenhuber, H. 1986. Technik und sozialer Wandel. In: Technik und sozialer Wandel. Ed: Lutz, B. Campus, Frankfurt.
Späth, L., Dallinger, P. 1998. Zehn Fragen an Lothar Späth. Das Hochschulwesen. 46: 75.
Späth, L. 1993. Eine Chance für zukunftsweisende Neuansätze. Wirtschaftsdienst 73: 64.
MPG-Archiv, ZA 143 Köhler, AO27.
Kaufmann, S.H. 1987. Possible role of helper and cytolytic T lymphocytes in antibacterial defense: conclusions based on a murine model of listeriosis. Rev. Infect. Dis. 9 Suppl 5:650.
Kaufmann, S.H. 1988. CD8+ T lymphocytes in intracellular microbial infections. Immunol. Today 9:168.
St, Johnston D., Nusslein-Volhard, C. 1992. The origin of pattern and polarity in the *Drosophila* embryo. Cell 68:201.
Nusslein-Volhard C. 1996. Gradients that organize embryo development. Sci. Am. 275:54; 58.
Shizuru JA, Negrin RS, Weissman IL. 2005. Hematopoietic stem and progenitor cells: Clinical and preclinical regeneration of the hematolymphoid system. Annu. Rev. Med. 56:509.
Illmensee K, Hoppe PC. 1981. Nuclear transplantation in Mus musculus: developmental potential of nuclei from preimplantation embryos. Cell 23:9.
Hoppe, P.C., Illmensee, K. 1982. Full-term development after transplantation of parthenogenetic embryonic nuclei into fertilized mouse eggs. Proc. Nat. Acad. Sci. USA 79:1912.
McGrath, J., Solter, D. 1983. Nuclear transplantation in the mouse embryo by microsurgery and cell fusion. Science 220:1300.
McGrath, J., Solter, D. 1984. Completion of mouse embryogenesis requires both the maternal and paternal genomes. Cell 37:179.
Solter D. 1988. Differential imprinting and expression of maternal and paternal genomes. Annu. Rev. Genet. 22:127.
University of Geneva: Karl Illmensee resigns. 1985. Nature 316: 98.
MPG-Archiv, ZA 143 Köhler, AO24-25, AO54, AO62.

## Chapter 12

MPG-Archiv, ZA 143 Köhler, AO1, AO3, AO12, AO32, AO34, AO55, AO56, AO62, AO66,
MPG-Archiv, II Abt., Rep. 1A PA Köhler.
Köhler, G., Eichmann, K., Eds. 1985. Immunsystem. Spektrum der Wissenschaft. Heidelberg.
Bäumler, E. 1984. Paul Ehrlich, scientist for life. Holmes and Meier Publ. New York, London.
Restany, P. 1992. Dani Karavan. Neues Pub. Co.

Karavan, D. 1995. "Passages" place of remembrance to Walter Benjamin at Portbou. Philipp von Zabern Verl.
Coutinho, A. 2003. Will the idiotypic network help to solve natural tolerance? Trends Immunol. 24:53.
Coutinho, A. 2000. Germ-line selection ensures embryonic autoreactivity and a positive discrimination of self mediated by supraclonal mechanisms. Semin. Immunol. 12:205.
Reimann, J., Schirmbeck, R. 1999. Alternative pathways for processing exogenous and endogenous antigens that can generate peptides for MHC class I-restricted presentation. Immunol. Rev. 172:131.
Zachau, H.G. 2000. The immunoglobulin kappa gene families of human and mouse: a cottage industry approach. Biol Chem. 381: 951.
Hämmerling, G.J., Schönrich, G., Ferber, I., Arnold, B. 1993.Peripheral tolerance as a multi-step mechanism. Immunol. Rev. 133:93.
Arnold, B., Schönrich, G., Hämmerling, GJ. 1993. Multiple levels of peripheral tolerance. Immunol. Today 14:12.
Reth, M.G., Alt, F.W. 1984. Novel immunoglobulin heavy chains are produced from DJH gene segment rearrangements in lymphoid cells. Nature 312:418.
Hombach, J., Leclercq, L., Radbruch, A., Rajewsky, K., Reth, M. 1988. A novel 34-kd protein co-isolated with the IgM molecule in surface IgM-expressing cells. EMBO J. 7:3451.
Reth M. 1989. Antigen receptor tail clue. Nature 338:383.
Reth, M., Wienands, J. 1999. The maintenance and the activation signal of the B-cell antigen receptor. Cold Spring Harb. Symp. Quant. Biol. 64:323.

## Chapter 13

Cotton, R.G.H., Milstein. C. 1973. Fusion of two immunoglobulin-producing myeloma cells. Nature 244: 42.
Grosschedl, R., Weaver, D., Baltimore, D., Costantini, F. 1984. Introduction of a mu immunoglobulin gene into the mouse germ line: specific expression in lymphoid cells and synthesis of functional antibody. Cell 38:647.
Storb, U., O'Brien, R.L., McMullen, M.D., Gollahon, K.A., Brinster, R.L. 1984. High expression of cloned immunoglobulin kappa gene in transgenic mice is restricted to B lymphocytes. Nature 310:238.
Ochi, A., Hawley, R.G., Hawley, T., Shulman, M.J., Traunecker, A., Köhler, G., Hozumi, N. 1983. Functional immunoglobulin M production after transfection of cloned immunoglobulin heavy and light chain genes into lymphoid cells. Proc. Nat. Acad. Sci. USA 80:6351.
Rusconi, S., Köhler, G. 1985. Transmission and expression of a specific pair of rearranged immunoglobulin mu and kappa genes in a transgenic mouse line. Nature 314:330.
Leder. P. 1982. The genetics of antibody diversity. Sci. Am. 246:108.
Kindt, T. Capra, J.D. 1984. The antibody enigma. Plenum Press, New York, London
Janeway, C.A.Jr., Travers, P., Walpot, M., Shlomchik, M.J. 2005. Immunobiology, the immune system in health and disease. 6th edition. GS Garland Science, New York, London.
Kottmann, A.H., Brack, C., Eibel, H., Köhler, G. 1992. A survey of protein-DNA interaction sites within the murine immunoglobulin heavy chain locus reveals a

particularly complex pattern around the DQ52 element. Eur. J. Immunol. 22:2113.

Kottmann, A,H,, Zevnik, B., Welte, M., Nielsen, P.J., Köhler G. 1994. A second promoter and enhancer element within the immunoglobulin heavy chain locus. Eur. J. Immunol. 24:817.

Iglesias, A., Lamers, M., Köhler, G. 1987. Expression of immunoglobulin delta chain causes allelic exclusion in transgenic mice. Nature 330:482.

Köhler, G. 1989. Allelic exclusion-one facet of B-cell development. Immunol. Today 10:11.

Melchers, F., ten Boekel, E., Yamagami, T., Andersson, J., Rolink, A. 1999. The roles of preB and B cell receptors in the stepwise allelic exclusion of mouse IgH and L chain gene loci. Semin. Immunol. 11:307

Loffert, D., Schaal, S., Ehlich, A., Hardy, R.R., Zou, Y.R., Muller, W., Rajewsky, K. 1994. Early B-cell development in the mouse: insights from mutations introduced by gene targeting. Immunol. Rev. 137:135.

Nitschke, L., Kosco, M.H., Köhler, G., Lamers, M.C. 1993. Immunoglobulin D-deficient mice can mount normal immune responses to thymus-independent and -dependent antigens. Proc. Nat. Acad. Sci. USA 90:1887.

Rammensee, H.G., Julius, M.H., Nemazee, D., Langhorne, J., Lamers, R., Köhler, G. 1987. Targeting cytotoxic T cells to antigen-specific B lymphocytes. Eur. J. Immunol. 17:433.

Julius, M.H., Rammensee, H.G., Ratcliffe, M.J., Lamers, M.C., Langhorne, J., Köhler, G. 1988. The molecular interactions with helper T cells which limit antigen-specific B cell differentiation. Eur. J. Immunol. 18:381.

Brombacher, F., Lamers, M.C., Köhler, G., Eibel, H. 1989. Elimination of CD8+ thymocytes in transgenic mice expressing an anti-Lyt2.2 immunoglobulin heavy chain gene. EMBO J. 8:3719.

Brombacher, F., Köhler, G., Eibel, H. 1991. B cell tolerance in mice transgenic for anti-CD8 immunoglobulin mu chain. J. Exp. Med. 174:1335.

Theopold, U., Köhler, G. 1990. Partial tolerance in beta-galactosidase-transgenic mice. Eur. J. Immunol. 20:1311.

Goodnow, C.C., Crosbie, J., Adelstein, S., Lavoie, T.B., Smith-Gill, S.J., Brink, R.A., Pritchard-Briscoe, H., Wotherspoon, J.S., Loblay, R.H., Raphael, K, et al. 1988. Altered immunoglobulin expression and functional silencing of self-reactive B lymphocytes in transgenic mice. Nature 334:676.

Nemazee, D.A., Burki, K. 1989. Clonal deletion of B lymphocytes in a transgenic mouse bearing anti-MHC class I antibody genes. Nature 337:562.

Götz, J., Eibel, H., Köhler, G. 1990. Non-tolerance and differential susceptibility to diabetes in transgenic mice expressing major histocompatibility class II genes on pancreatic beta cells. Eur. J. Immunol. 20:1677.

Allison, J., Campbell, I.L., Morahan, G., Mandel, T.E., Harrison, L.C., Miller, J.F. 1988. Diabetes in transgenic mice resulting from over-expression of class I histocompatibility molecules in pancreatic beta cells. Nature 333:529.

Lo, D., Burkly, L.C., Widera, G., Cowing, C., Flavell, R.A., Palmiter, R.D., Brinster, R.L. 1988. Diabetes and tolerance in transgenic mice expressing class II MHC moleculein pancreatic beta cells. Cell53:159.

Sarvetnick, N., Liggitt, D., Pitts, S.L., Hansen, S.E., Stewart, T.A. 1988. Insulin-dependent diabetes mellitus induced in transgenic mice by ectopic expression of class II MHC and interferon-gamma. Cell 52:773.

Bohme, J., Haskins, K., Stecha, P., van Ewijk, W., LeMeur M., Gerlinger, P., Benoist,

C., Mathis, D. 1989. Transgenic mice with I-A on islet cells are normoglycemic but immunologically intolerant. Science 244:1179.

## Chapter 14

Jacob, F. 1978. The Leeuwenhoek Lecture, 1977. Mouse teratocarcinoma and mouse embryo. Proc. R. Soc. Lond. B Biol. Sci. 201:249.

Evans, M.J., Kaufman, M.H. 1981. Establishment in culture of pluripotential cells from mouse embryos. Nature 292:154.

Kemler, R., Babinet, C., Eisen, H., Jacob, F. 1977. Surface antigen in early differentiation. Proc. Nat. Acad. Sci. USA 74:4449.

Doetschman, T.C., Eistetter, H., Katz, M., Schmidt, W., Kemler, R. 1985. The in vitro development of blastocyst-derived embryonic stem cell lines: formation of visceral yolk sac, blood islands and myocardium. J. Embryol. Exp. Morphol. 87:27.

Bradley, A., Evans, M., Kaufman, M.H., Robertson, E. 1984. Formation of germ-line chimaeras from embryo-derived teratocarcinoma cell lines. Nature 309:255.

Gossler, A., Doetschman, T., Korn, R., Serfling, E., Kemler, R. 1986. Transgenesis by means of blastocyst-derived embryonic stem cell lines. Proc. Nat. Acad. Sci. USA 83:9065.

Doetschman, T., Gregg, R.G., Maeda, N., Hooper, M.L., Melton, D.W., Thompson, S., Smithies, O. 1987. Targetted correction of a mutant HPRT gene in mouse embryonic stem cells. Nature 330:576.

Thomas K.R., Capecchi, M.R. 1987. Site-directed mutagenesis by gene targeting in mouse embryo-derived stem cells. Cell 51:503.

Mansour, S.L., Thomas, K.R., Capecchi, M.R. 1988. Disruption of the proto-oncogene int-2 in mouse embryo-derived stem cells: a general strategy for targeting mutations to non-selectable genes. Nature 336:348.

Zijlstra, M., Bix, M., Simister, N.E., Loring, J.M., Raulet, D.H., Jaenisch, R. 1990. Beta 2-microglobulin deficient mice lack CD4-8+ cytolytic T cells. Nature 344:742.

Huber, O., Bierkamp, C., Kemler, R. 1996. Cadherins and catenins in development. Curr. Opin. Cell. Biol. 8:685.

Aoki, M., Hecht, A., Kruse, U., Kemler, R., Vogt, P.K. 1999. Nuclear endpoint of Wnt signaling: neoplastic transformation induced by transactivating lymphoid-enhancing factor 1. Proc. Nat. Acad. Sci. USA 96:139.

Kitamura, D., Roes, J., Kuhn, R., Rajewsky, K. 1991. A B cell-deficient mouse by targeted disruption of the membrane exon of the immunoglobulin mu chain gene. Nature 350:423.

Rajewsky, K., Gu, H., Kuhn, R., Betz, U.A., Muller, W., Roes, J, Schwenk, F. 1996. Conditional gene targeting. J. Clin. Invest. 98:600.

Kuhn, R., Schwenk, F., Aguet, M., Rajewsky, K. 1995. Inducible gene targeting in mice. Science 269:1427.

Coffman, R.L, Mosmann, T.R. 1991. CD4+ T-cell subsets: regulation of differentiation and function. Res. Immunol. 142:7.

Kuhn, R., Rajewsky, K., Müller, W. 1991. Generation and analysis of interleukin-4 deficient mice. Science 254:707.

Kopf, M., Le Gros, G., Bachmann, M., Lamers, M.C., Bluethmann, H., Köhler, G. 1993. Disruption of the murine IL-4 gene blocks Th2 cytokine responses. Nature 362:245.

von der Weid, T., Kopf, M., Köhler, G., Langhorne, J. 1994. The immune response to

*Plasmodium chabaudi* malaria in interleukin-4-deficient mice. Eur. J. Immunol. 24:2285.

Kopf, M., Brombacher, F., Köhler, G., Kienzle, G., Widmann, K.H., Lefrang, K., Humborg, C., Ledermann, B., Solbach, W. 1996. IL-4-deficient Balb/c mice resist infection with *Leishmania major*. J. Exp. Med. 184:1127.

Noben-Trauth, N., Kropf, P., Müller, I. 1996. Susceptibility to *Leishmania major* infection in interleukin-4-deficient mice. Science 271:987.

Kopf, M., Baumann, H., Freer, G., Freudenberg, M., Lamers, M., Kishimoto, T., Zinkernagel, R., Bluethmann, H., Köhler, G. 1994. Impaired immune and acute-phase responses in interleukin-6-deficient mice. Nature 368:339.

Ramsay, A.J., Husband, A.J., Ramshaw, I.A., Bao, S., Matthaei, K.I., Köhler, G., Kopf, M. 1994. The role of interleukin-6 in mucosal IgA antibody responses *in vivo*. Science. 264:561.

Bernad, A., Kopf, M., Kulbacki., R, Weich, N., Köhler, G., Gutierrez-Ramos, J.C. 1994. Interleukin-6 is required *in vivo* for the regulation of stem cells and committed progenitors of the hematopoietic system. Immunity. 1:725.

Kopf, M., Le Gros, G., Coyle, A.J., Kosco-Vilbois, M., Brombacher, F. 1995. Immune responses of IL-4, IL-5, IL-6 deficient mice. Immunol. Rev. 148:45.

Nielsen, P.J., Lorenz, B., Muller, A.M., Wenger, R.H., Brombacher, F., Simon, M., von der Weid, T., Langhorne, J., Mossmann, H., Köhler, G. 1997. Altered erythrocytes and a leaky block in B-cell development in CD24/HSA-deficient mice. Blood. 89:1058.

## Chapter 15

Deutsches Museum Bonn, AO Monoclonale Antikörper.

Köhler, F., Storch, B., Kulathu, Y., Herzog, S., Kuppig, S., Reth, M., Jumaa, H. 2005. A leucine zipper in the N terminus confers membrane association to SLP-65. Nat. Immunol. 6:204.

## Chapter 16

Isaacs, A., Lindenmann, J. 1987 (reprinted). Virus interference. I. The interferon. 1957. J Interferon Res. 7:429.

Peters, T. 1993. Interferon and its first clinical trial: Looking behind the scenes. Medical History 37: 270.

Secher, D.S., Burke, D.C. 1980. A monoclonal antibody for large-scale purification of human leukocyte interferon. Nature 285:446.

Mountain, A., Ney, U., Schomburg, D. Eds. 1999. Biotechnology 2nd Ed. Volume 5a. Recombinant proteins, monoclonal antibodies, and therapeutic genes. Wiley-VCH. Weinheim.

Bullok, D. 1983. Diagnostic and biochemical applications: Monoclonal antibodies. Biotechnology Advances 84: 81.

Payne, W.J., Marshall, D.L., Shockley, R.K., Martin, W.J. 1988. Clinical laboratory applications of monoclonal antibodies. J. Immunol. Meth. 116: 151.

Chein, S., Silverstein, S.C. 1993. Economic impact of application of monoclonal antobodies to medicine and biology. The FASEB J. 7:1426.

World market for monoclonal anitibodies 2001-2010. 2004. TriMark Publications, New York. http://trimarkpublications.ecnext.com

Monoclonal antibodies 2003: Meeting clinical and financial expectations. 2004. Lead discovery services, UK. http://www.leaddiscovery. co.uk.

Fjermedal. 1984. Magic bullets. Macmillan, New York.

Berkower, I. 1996. The promose and pitfalls of monoclonal antibody therapeutics. Curr. Op. Biotech. 7:622.

Winter, G., Milstein, C. 1991. Man-made antibodies. Nature 349:293.

Morrison, S.L. 1992. *In vitro* antibodies: Strategies for production and application. Ann. Rev. Immunol. 10:239.

Grillo-Lopez, J. Ed. 2001. Monoclonal antibodies in the treatment of hematologic malignancies. In: Current Pharmaceutical Biotechnology, Vol.2. Bentham Science Publ.

Weiner, G.J., Link, B.K. 2004. Antibody therapy of lymphoma. Adv. Pharmacol. 51:229.

Knight, C., Hind, D., Brewer, N., Abbott, V. 2004. Rituximab (MabThera) for aggressive non-Hodgkin's lymphoma: systematic review and economic evaluation. Health Technol. Assess. 8: 1.

Hillmen, P. 2004. Advancing therapy for chronic lymphocytic leukemia-the role of rituximab. Semin. Oncol. 31(1 Suppl 2):22.

Dechant, M., Bruenke, J., Valerius, T. 2003. HLA class II antibodies in the treatment of hematologic malignancies. Semin. Oncol. 30:465.

Ghobrial, I., Witzig, T. 2004. Radioimmunotherapy: a new treatment modality for B-cell non-Hodgkin's lymphoma. Oncology 18:623.

Linenberger, M.L. 2005. CD33-directed therapy with gemtuzumab ozogamicin in acute myeloid leukemia: progress in understanding cytotoxicity and potential mechanisms of drug resistance. Leukemia19:176.

Ravandi, F., O'Brien, S. 2005. Alemtuzumab. Expert Rev. Anticancer Ther. 5:39.

Dearden, C. 2004. Alemtuzumab in peripheral T-cell malignancies. Cancer Biother. Radiopharm. 19:391.

Kataoka, A., Ishidam, M., Murakamim, S., Ohnom S. 2004. Sensitization of chemotherapy by anti-HER. Breast Cancer 11:105.

Bell , R., Verma, S., Untch, M., Cameron, D., Smith, I. 2004. Maximizing clinical benefit with trastuzumab. Semin. Oncol. 31(5 Suppl 10):35.

Emens, L.A., Davidson, N.E. 2004. Trastuzumab in breast cancer. Oncology 18:1117.

Ng, M., Cunningham, D. 2004. Cetuximab (Erbitux) – an emerging targeted therapy for epidermal growth factor receptor-expressing tumours. Int. J. Clin. Pract. 58:970.

Baumann, M., Krause, M. 2004. Targeting the epidermal growth factor receptor in radiotherapy: radiobiological mechanisms, preclinical and clinical results. Radiother. Oncol. 72:257.

Camp-Sorrell, D. 2003. Antiangiogenesis: the fifth cancer treatment modality? Oncol. Nurs. Forum. 30:934.

Glade-Bender, J., Kandel, J.J., Yamashiro, D.J. 2003. VEGF blocking therapy in the treatment of cancer. Expert Opin. Biol. Ther. 3:263.

Franson, P.J., Lapka, D.V. 2005. Antivascular endothelial growth factor monoclonal antibody therapy: a promising paradigm in colorectal cancer. Clin. J, Oncol. Nurs. 9:55.

Iqbal, S., Lenz. H-J. 2004. Angiogenesis inhibitors in the treatment of colorectal cancer. Semin. Oncol. 31(6 Suppl 17):10.

Roberts, L., McColl, G.J. 2004. Tumour necrosis factor inhibitors: risks and benefits in patients with rheumatoid arthritis. Intern. Med. J. 34:687.

Sands, B.E. 2004. Why do anti-tumor necrosis factor antibodies work in Crohn's disease? Rev. Gastroenterol Disord. 4 (Suppl 3):10.
Ehlers, S. 2005. Why does tumor necrosis factor targeted therapy reactivate tuberculosis? J. Rheumatol. 74:35.
Buhl, R. 2005. Anti-IgE antibodies for the treatment of asthma. Curr. Opin. Pulm. Med. 11:27.
Bang, L.M., Plosker, G.L. 2004. Spotlight on omalizumab in allergic asthma. BioDrugs 18:415.
Van Gelder, T., Warle, M., Ter Meulen, R.G. 2004. Anti-interleukin-2 receptor antibodies in transplantation: what is the basis for choice? Drugs 64:1737.

## Chapter 17

Maizels, N. 1995. Somatic hypermutation: how many mechanisms diversify V region sequences? Cell. 83:9.
Wagner, S.D., Neuberger, M.S. 1996. Somatic hypermutation of immunoglobulin genes. Annu. Rev. Immunol. 14:441.
Muramatsu, M., Kinoshita, K., Fagarasan, S., Yamada, S., Shinkai, Y., Honjo, T. 2000. Class switch recombination and hypermutation require activation-induced cytidine deaminase (AID), a potential RNA editing enzyme. Cell 102:553.
Petersen-Mahrt, S.K., Harris, R.S., Neuberger, M.S. 2002. AID mutates *E. coli* suggesting a DNA deamination mechanism for antibody diversification. Nature 418:99.
Kinoshita, K., Honjo, T. 2001. Linking class-switch recombination with somatic hypermutation. Nat. Rev. Mol. Cell Biol. 2:493.
Honjo, T., Muramatsu, M., Fagarasan, S. 2004. AID: how does it aid antibody diversity? Immunity 20:659.
Neuberger, M.S., Harris, R.S., Di Noia, J., Petersen-Mahrt, S.K. 2003. Immunity through DNA deamination. Trends Biochem. Sci. 28:305.
Neuberger, M.S., Di Noia, J.M., Beale, R.C., Williams, G.T., Yang, Z., Rada, C. 2005. Somatic hypermutation at A.T pairs: polymerase error versus dUTP incorporation. Nat. Rev. Immunol. 5:171.

# Köhler's complete bibliography

Köhler, G., Melchers, F. 1972. Isoelectric focusing spectra of antibodies which activate mutant b-galactosidases. Eur. J. Immunol. 2:453.

Köhler, G., Milstein, C. 1975. Continuous cultures of fused cells secreting antibody of predefined specificity. Nature 256:495.

Köhler, G., Howe, S.C., Milstein, C. 1976. Fusion between immunoglobulin-secreting and nonsecreting myeloma cell lines. Eur. J. Immunol. 6:292.

Ziegler, A., Köhler, G. 1976. Analytical isoelectric focusing in polymerizable thin layers containing Sephadex. FEBS Lett. 64:48.

Köhler, G. 1976. Frequency of precursor cells against the enzyme beta-galactosidase: an estimate of the BALB/c strain antibody repertoire. Eur. J. Immunol. 6:340.

Köhler, G., Milstein, C. 1976. Derivation of specific antibody-producing tissue culture and tumor lines by cell fusion. Eur. J. Immunol. 6:511.

Ziegler, A., Köhler, G. 1976. Resolving power of isotachophoresis and isoelectric focusing for immunoglobulins. FEBS Lett. 72:142.

Milstein, C., Adetugbo, K., Cowan, N.J., Köhler, G., Secher, D.S., Wilde, C.D. 1977. Somatic cell genetics of antibody-secreting cells: studies of clonal diversification and analysis by cell fusion. Cold Spring Harb. Symp. Quant. Biol. 41 Pt 2:793.

Köhler, G., Pearson, T., Milstein, C. 1977. Fusion of T and B cells. Somatic Cell Genet. 3:303.

Köhler, G., Lefkovits, I., Elliott, B., Coutinho, A. 1977. Derivation of hybrids between a thymoma line and spleen cells activated in a mixed leukocyte reaction. Eur. J. Immunol. 7:758.

Köhler, G., Shulman, M.J. 1978. Cellular and molecular restrictions of the lymphocyte fusion. Curr. Top. Microbiol. Immunol. 81:143.

Köhler, G., Hengartner, H., Shulman, M.J. 1978. Immunoglobulin production by lymphocyte hybridomas. Eur. J. Immunol. 8:82.

Milstein, C., Adetugbo, K., Cowan, N.J., Köhler, G,. Secher, D.S. 1978. Expression of antibody genes in tissue culture: structural mutants and hybrid cells. Natl. Cancer Inst. Monogr. 48:321.

Shulman, M.J., Köhler, G. 1978. Immunoglobulin mu and gamma heavy chain classes are not determined by class-specific RNA-splicing enzymes. Nature 274:917.

Shulman, M., Wilde, C.D., Köhler, G. 1978. A better cell line for making hybridomas secreting specific antibodies. Nature 276:269.

Forni, L., Coutinho, A., Köhler, G., Jerne, N.K. 1980. IgM antibodies induce the production of antibodies of the same specificity. Proc. Natl. Acad. Sci. USA 77:1125.

Köhler, G. 1980. Immunoglobulin chain loss in hybridoma lines. Proc. Natl. Acad. Sci. USA 77:2197.

Notenboom, R.H., Chou, C.T., Good, P.W., Dubiski, S., Cinader, B., Köhler, G. 1980. Isolation and characterization of a mouse-rabbit hybridoma. J. Immunogenet. 7:359.

Köhler, G,1981. Why hybridomas? Hybridoma 1:1.

Sidman, C., Potash, M.J., Köhler, G. 1981. Roles of protein and carbohydrate in glycoprotein processing and secretion. Studies using mutants expressing altered IgM mu chains. J. Biol. Chem. 256:13180.

Köhler, G., Potash, M.J., Lehrach, H., Shulman, M.J. 1982. Deletions in immunoglobulin mu chains. EMBO J. 1:555.

Hirayama, N., Hirano,T., Köhler, G., Kurata, A., Okumura, K., Ovary, Z. 1982. Biological activities of antitrinitrophenyl and antidinitrophenyl mouse monoclonal antibodies. Proc. Natl. Acad. Sci. USA 79:613.

Shulman, M.J., Heusser, C., Filkin, C., Köhler, G. 1982. Mutations affecting the structure and function of immunoglobulin M. Mol. Cell. Biol. 2:1033.

Sidman, C.L., Forni, L., Köhler, G., Langhorne, J., Lindahl, K.F. 1983. A monoclonal antibody against a new differentiation antigen of thymocytes. Eur. J. Immunol. 13:481.

Ochi, A., Hawley, R.G., Hawley, T., Shulman, M.J., Traunecker, A., Köhler, G., Hozumi, N. 1983. Functional immunoglobulin M production after transfection of cloned immunoglobulin heavy and light chain genes into lymphoid cells. Proc. Natl. Acad.Sci. USA 80:6351.

Köhler, G., Baumann, B., Iglesias, A., McCubrey, J., Potash, M.J., Traunecker, A., Zhu, D. 1984. Different ways to modify monoclonal antibodies. Med. Oncol. Tumor Pharmacother. 1:227.

Shulman, M.J., Hawley, R.G., Ochi, A., Baczynsky, W.O., Collins, C., Pennell, N., Potash. M.J., Köhler, G., Hozumi, N. 1984. Biochemical genetics of the mouse IgM system. Can. J. Biochem. Cell. Biol. 62:217.

Leptin, M., Potash, M.J., Grutzmann, R., Heusser, C., Shulman, M., Köhler, G., Melchers, F. 1984. Monoclonal antibodies specific for murine IgM I. Characterization of antigenic determinants on the four constant domains of the mu heavy chain. Eur. J. Immunol. 14:534.

Zhu, D., Lefkovits, I., Köhler, G. 1984. Frequency of expressed immunoglobulin light chain genes in lipopolysaccharide-stimulated BALB/c spleen cells. J. Exp. Med. 160:971.

Baumann, B., Potash, M.J., Köhler, G. 1985. Consequences of frameshift mutations at the immunoglobulin heavy chain locus of the mouse. EMBO J. 4:351.

Rusconi, S., Köhler, G. 1985. Transmission and expression of a specific pair of rearranged immunoglobulin mu and kappa genes in a transgenic mouse line. Nature 314:330.

Köhler, G. 1985. Derivation and diversification of monoclonal antibodies. EMBO J. 4:1359.

Köhler G. 1985. Derivation and diversification of monoclonal antibodies. Nobel lecture, 8 December 1984. Biosci. Rep. 5:533.

McCubrey, J., McKearn, J.P., Köhler, G. 1985. Transformation of B and non-B cell lines with the 2,4,6,-trinitrophenyl (TNP)-specific immunoglobulin genes. Eur. J. Immunol. 15:1117.

Köhler, G. 1986. Derivation and diversification of monoclonal antibodies. Science 233:1281.

Rammensee, H.G., Julius, M.H., Nemazee, D., Langhorne, J., Lamers, R., Köhler, G. 1987. Targeting cytotoxic T cells to antigen-specific B lymphocytes. Eur. J. Immunol. 17:433.

Iglesias, A., Lamers, M., Köhler, G. 1987. Expression of immunoglobulin delta chain causes allelic exclusion in transgenic mice. Nature 330:482.

Mathur, A,, Lynch, R.G., Köhler, G. 1988. The contribution of constant region domains to the binding of murine IgM to Fc mu receptors on T cells. J. Immunol. 140:143.

Julius, M.H., Rammensee, H.G., Ratcliffe, M.J., Lamers, M.C., Langhorne, J., Köhler, G. 1988. The molecular interactions with helper T cells which limit antigen-specific B cell differentiation. Eur. J. Immunol. 18:381.

Mathur, A., Lynch, R.G., Köhler, G. 1988. Expression, distribution and specificity of Fc receptors for IgM on murine B cells. J. Immunol. 141:1855.

Lamers, M.C., Vakil, M., Kearney, J.F., Langhorne, J., Paige, C.J., Julius, M.H., Mossmann, H., Carsetti, R., Köhler G. 1989. Immune status of a mu, kappa transgenic mouse line. Deficient response to bacterially related antigens. Eur. J. Immunol. 19:459.

Köhler, G. 1989. Allelic exclusion--one facet of B-cell development. Immunol. Today10:11.

Ayane, M., Nielsen, P., Köhler, G. 1989. Cloning and sequencing of mouse ribosomal protein S12 cDNA. Nucleic Acids Res. 17:6722.

Gutierrez-Ramos, J.C., Martinez, C., Köhler, G., Iglesias, A. 1989. Analysis of T-cell subpopulations in human IL-2R alpha transgenic mice: expansion of Thy1.2- thymocytes and depletion of double-positive T-cell precursors. Res. Immunol. 140:661.

Brombacher, F., Lamers, M.C., Köhler, G., Eibel, H. 1989. Elimination of CD8+ thymocytes in transgenic mice expressing an anti-Lyt2.2 immunoglobulin heavy chain gene. EMBO J. 8:3719.

Theopold, U., Köhler, G. 1990. Partial tolerance in beta-galactosidase-transgenic mice. Eur. J. Immunol. 20:1311.

Gotz, J., Eibel, H., Köhler, G. 1990. Non-tolerance and differential susceptibility to diabetes in transgenic mice expressing major histocompatibility class II genes on pancreatic beta cells. Eur. J. Immunol. 20:1677.

Adamczewski, M., Köhler, G., Lamers, M.C. 1991. Expression and biological effects of high levels of serum IgE in epsilon heavy chain transgenic mice. Eur. J. Immunol. 21:617.

Ayane, M., Preuss, U., Köhler, G., Nielsen, P.J. 1991. A differentially expressed murine RNA encoding a protein with similarities to two types of nucleic acid binding motifs. Nucleic. Acids. Res. 19:1273.

Wenger, R.H., Ayane, M., Bose, R., Köhler, G., Nielsen, P.J. 1991. The genes for a mouse hematopoietic differentiation marker called the heat-stable antigen. Eur. J. Immunol. 21:1039.

Iglesias, A., Kopf, M., Williams, G.S., Buhler, B., Köhler G. 1991. Molecular requirements for the mu-induced light chain gene rearrangement in pre-B cells. EMBO J. 10:2147.

Brombacher, F., Köhler, G., Eibel, H. 1991. B cell tolerance in mice transgenic for anti-CD8 immunoglobulin mu chain. J. Exp. Med. 174:1335.

Eibel, H., Brombacher, F., Köhler, G. 1992. Analysis of B-cell tolerance in mice expressing transgenic anti-CD8.2 immunoglobulin M molecules. Res Immunol. 143:276.

Kottmann, A.H., Brack, C., Eibel, H., Köhler G. 1992. A survey of protein-DNA interaction sites within the murine immunoglobulin heavy chain locus reveals a particularly complex pattern around the DQ52 element. Eur. J. Immunol. 22:2113.

Carsetti, R., Köhler, G., Lamers, M.C. A role for immunoglobulin D: interference with tolerance induction. Eur. J. Immunol. 23:168.

Nitschke, L., Kosco, M.H., Köhler, G., Lamers, M.C. 1993. Immunoglobulin D-deficient mice can mount normal immune responses to thymus-independent and -dependent antigens. Proc. Natl. Acad. Sci. USA 90:1887.

Kopf, M., Le Gros, G., Bachmann, M., Lamers, M.C., Bluethmann, H., Köhler, G. 1993. Disruption of the murine IL-4 gene blocks Th2 cytokine responses. Nature 362:245.

Iglesias, A., Nichogiannopoulou, A., Williams, G.S., Flaswinkel, H., Köhler, G. 1993. Early B cell development requires mu signaling. Eur. J .Immunol. 23:2622.

Nielsen, P.J., Eichmann, K., Köhler, G., Iglesias, A. 1993. Constitutive expression of transgenic heat stable antigen (mCD24) in lymphocytes can augment a secondary antibody response. Int. Immunol. 5:1355.

Wenger, R.H., Rochelle, J.M., Seldin, M.F., Köhler, G., Nielsen, P.J. 1993. The heat stable antigen (mouse CD24) gene is differentially regulated but has a housekeeping promoter. J. Biol. Chem. 268:23345.

Dirkes, G,. Köhler, G., Kottmann, A.H. 1994. Sequence and structure of the mouse IgH DQ52 5' region. Immunogenetics 40:379.

Kopf, M., Baumann, H., Freer, G., Freudenberg, M., Lamers, M., Kishimoto, T., Zinkernagel, R., Bluethmann, H., Köhler, G. 1994. Impaired immune and acute-phase responses in interleukin-6-deficient mice. Nature. 368:339.

Kottmann, A.H., Zevnik, B., Welte, M., Nielsen, P.J., Köhler, G. 1994. A second promoter and enhancer element within the immunoglobulin heavy chain locus. Eur. J. Immunol. 24:817.

Yu, P., Kosco-Vilbois, M., Richards, M., Köhler, G., Lamers, M.C. 1994. Negative feedback regulation of IgE synthesis by murine CD23. Nature 369:753.

Terashima, M., Kim, K.M., Adachi, T., Nielsen, P.J., Reth, M., Köhler, G., Lamers, M.C. 1994. The IgM antigen receptor of B lymphocytes is associated with prohibitin and a prohibitin-related protein. EMBO J. 13:3782 .

Kim, K.M., Adachi, T., Nielsen ,P.J., Terashima, M., Lamers, M.C., Köhler, G., Reth, M. 1994. Two new proteins preferentially associated with membrane immunoglobulin D. EMBO J. 13:3793.

von der Weid, T., Kopf, M., Köhler, G., Langhorne, J. 1994. The immune response to Plasmodium chabaudi malaria in interleukin-4-deficient mice. Eur. J. Immunol. 24:2285.

Buhler, B., Köhler, G., Nielsen, P.J. 1994. A lacZ-based vector system for the rapid detection of V(D)J recombinase activity. J. Immunol. Methods. 175:259.

Buhler, B., Köhler, G,, Nielsen, P.J. 1995. Efficient nonhomologous and homologous recombination in scid cells. Immunogenetics. 42:181.

Vajdy, M,, Kosco-Vilbois, M.H., Kopf, M., Köhler, G., Lycke, N. 1995. Impaired mucosal immune responses in interleukin 4-targeted mice. J. Exp. Med. 181:41.

Wenger, R.H., Kopf, M., Nitschke, L., Lamers, M.C., Köhler. G,, Nielsen. P.J. 1995. B-cell maturation in chimaeric mice deficient for the heat stable antigen (HSA/mouse CD24). Transgenic Res. 4:173.

Carsetti, R., Köhler, G., Lamers, M.C. 1995. Transitional B cells are the target of negative selection in the B cell compartment. J. Exp- Med. 181:2129.

Hilbert, D.M., Kopf, M., Mock, B.A., Köhler, G., Rudikoff, S. 1995. Interleukin 6 is essential for *in vivo* development of B lineage neoplasms. J. Exp. Med. 182:243.

Kopf, M., Ramsay, A., Brombacher, F., Baumann, H., Freer, G., Galanos, C., Gutierrez-Ramos, J.C., Köhler G. 1995. Pleiotropic defects of IL-6-deficient mice including early hematopoiesis, T and B cell function, and acute phase responses. Ann. N. Y. Acad. Sci. 762:308.

Battegay, M., Fiedler, P., Kalinke, U., Brombacher, F., Zinkernagel, R.M,. Peter, H.H., Köhler, G., Eibel, H. 1996. Non-tolerant B cells cause autoimmunity in anti-CD8 IgG2a-transgenic mice. Eur. J. Immunol. 26:250.

Kopf, M., Brombacher, F., Hodgkin, P.D., Ramsay, A.J., Milbourne, E.A,. Dai, W.J., Ovington, K.S., Behm, C.A., Köhler, G., Young, I.G., Matthaei, K.I. 1996. IL-5-deficient mice have a developmental defect in CD5+ B-1 cells and lack eosinophilia but have normal antibody and cytotoxic T cell responses. Immunity 4:15.

Pearce, E.J., Cheever, A., Leonard, S., Covalesky, M., Fernandez-Botran, R., Köhler, G., Kopf, M. 1996. Schistosoma mansoni in IL-4-deficient mice. Int. Immunol. 8:435.

Scott, D.W., Lamers, M., Köhler, G., Sidman, C.L., Maddox, B., Carsetti, R. 1996. Role of c-myc and CD45 in spontaneous and anti-receptor-induced apoptosis in adult murine B cells. Int. Immunol. 8:1375.

Kopf, M., Brombacher, F., Köhler, G., Kienzle, G., Widmann, K.H., Lefrang, K., Humborg, C., Ledermann, B., Solbach, W. 1996. IL-4-deficient Balb/c mice resist infection with Leishmania major. J. Exp. Med. 184:1127.

Noben-Trauth, N., Köhler, G., Burki, K., Ledermann, B. 1996. Efficient targeting of the IL-4 gene in a BALB/c embryonic stem cell line. Transgenic Res. 5:487.

Nitschke, L., Carsetti, R., Ocker, B., Köhler, G., Lamers, M.C. 1997. CD22 is a negative regulator of B-cell receptor signalling. Curr. Biol. 7:133.

Nielsen, P.J., Lorenz, B., Muller, A.M., Wenger, R.H., Brombacher, F., Simon, M., von der Weid, T., Langhorne, W.J., Mossmann, H., Köhler. G. 1997. Altered erythrocytes and a leaky block in B-cell development in CD24/HSA-deficient mice. Blood 89:1058.

Lutz, C., Ledermann, B., Kosco-Vilbois, M.H., Ochsenbein, A.F., Zinkernagel, R.M., Köhler, G., Brombacher, F. 1998. IgD can largely substitute for loss of IgM function in B cells. Nature 393:797.

MIX
Papier aus verantwortungsvollen Quellen
Paper from responsible sources
FSC® C105338

If you have any concerns about our products,
you can contact us on
**ProductSafety@springernature.com**

In case Publisher is established outside the EU,
the EU authorized representative is:
**Springer Nature Customer Service Center GmbH**
**Europaplatz 3, 69115 Heidelberg, Germany**

Printed by Libri Plureos GmbH
in Hamburg, Germany